LASER LADY MEETS THE LIGHT JUNKIES

LASER LADY MEETS THE LIGHT JUNKIES

A HOLLYWOOD MEMOIR

LINDA LANE

Laser Lady Meets the Light Junkies: A Hollywood Memoir

This edition published by Highpoint LIT, an imprint of Highpoint Executive Publishing.

For information, write to info@highpointpubs.com.

First Edition
ISBN: 978-0-9796900-6-8

Library of Congress Cataloging-in-Publication Data

Lane, Linda
Laser Lady Meets the Light Junkies

Summary: "*Laser Lady Meets the Light Junkies* presents a history of early holography and a personal coming-of-age memoir in which a smart, visionary, ambitious, but somewhat naïve twentysomething woman strives to find success in a swirling pool of crazy scientists, backstabbing lawyers, Hollywood sharks, and assorted hustlers in the anything-goes 1970s." Provided by publisher.

ISBN: 978-0-9796900-6-8 (hardcover)
1. Memoir

Library of Congress Control Number: 2020910327

Cover and interior design by Sarah M. Clarehart

10 9 8 7 6 5 4 3 2 1

CONTENTS

DEDICATION

To Lucy and Philip

PREFACE

"Keep a diary when you're young and it will keep you when you're old."

—Margot Asquith, 1922

There are a number of compelling reasons for writing this memoir. First, and foremost, I survived the genesis of holography. I was in the right place at the right moment to witness and be part of a technological breakthrough that is now part of our daily lives: the hologram on your credit card, the holographic strip on a hundred dollar bill, Princess Leia delivering a message in *Star Wars*, or Boeing using a three-dimensional image of an airplane part to detect a flaw that would be imperceptible to the naked eye.

Seeing my first hologram in 1972 changed my life forever. I was no longer a confused college graduate. In that first moment, I knew exactly what I wanted to do. I came up with the idea for a screenplay, got the money to write it, and set out to find the visionaries and physicists who could make a larger-than-life-sized hologram for first-run theaters.

Guided by my daily diaries, I began writing a brief history of holography. Only within those pages I discovered something more. I came face-to-face with my younger self, and I was forced to relive the good, the bad, and the ugly—the high-highs and the low-lows. I was forced to come to terms with my decision-making processes. Why was I so naïve? So trusting? So ruthlessly optimistic? How did becoming "Laser Lady" motivate me to close my eyes and leap down the rabbit hole? These are all questions I needed to answer while writing this memoir.

The best way to introduce myself is to take you to Brentwood, a lovely suburb of Los Angeles, where in the 1950s and '60s, movie star tour bus drivers pointed bullhorns at my house, erroneously telling tourists that Linda Darnell, "the actress with the perfect face," lived there. I, of course, hated it, but secretly, those encounters made me feel special. I understood that Joan Crawford and Barbara Stanwyck lived down the street, across from one another. Novelist Sydney Sheldon was our next-door neighbor.

I would close my eyes and dream of my own *Architectural Digest*-perfect house with a view of the city, and the man of my dreams pulling into the driveway in his white Cadillac convertible to sweep me away to a happy new life.

Fantasy, as I now understand, is a useful survival mechanism. As an only child, I spent a lot of time alone. In those days, we had to amuse ourselves, and that meant developing creative skills. As early as middle school, I began keeping a daily diary. This allowed me to express my thoughts, feelings, and dreams in a criticism-free space. The diary couldn't argue with me, tell me I was too fat or that my legs weren't long enough, or that I talked too much. My diary was Switzerland, and I was safe.

In high school I kept a diary in French so that my mother couldn't read it. I lived in a large Tudor house with a bedroom the size of a single apartment. I slept in an antique Victorian queen-sized bed on professionally pressed sheets. There was a matching Victorian bureau with a mirror, and another chest of drawers with another mirror. The bookcase was filled with plays, including most of Shakespeare, and a tall cherry-wood desk hid the remnants of school projects. An antique armchair covered in bright pink velvet faced a wall of windows that overlooked the front garden with its sky-high pine tree, circular driveway lined with rose bushes, and the tall palm trees that bordered the distant street.

My bathroom sported pink wallpaper with a pink bathtub and shower. And, of course, I had a skylight. To the outside world it appeared that I had everything—everything, that is, except privacy. When I was on the phone upstairs, my mother would quietly lift a downstairs receiver and listen. She had begun to live vicariously through me.

My father never noticed anything. He owned a construction company that specialized in commercial buildings, and when he wasn't at his office, he was hunting or fishing. As the times prescribed, he followed a routine.

He left the house in the morning, and upon arriving back for dinner, poured himself a Scotch, ate a couple bites of cheddar cheese on a Ritz cracker, and watched the nightly news. When I was a child he designed a midnight blue fiberglass sports car that looked like a predecessor to the XKE Jaguar. He reeled in a world record-breaking 557-pound black sea bass on 45-pound tackle off Catalina Island, and he'd been California State Skeet Champion. He was over six feet tall and built like Thor. In short, to me he was larger-than-life, a man whose approval I desperately sought.

This was no easy task since I always seemed to be competing with my mother. From my birth, I suspect, she had vied with me for my father's attention. When I was five years old, he would empty his pockets, presenting me with a few coins before heading to his office, my mother would scream, "Stop giving her money. It's not right!"

This, of course, had nothing to do with the money. My father found it very difficult to show warmth or love. This was his way of bonding, of showing affection and sharing a moment with me. My mother saw it as her daughter stealing her husband's affection, and the older I got the more crushing her accusations became.

My parents were glamorous, and on the face of it they had everything anyone could want. In my early years they behaved well in public; however, behind closed doors, all of their frustration and discontent would bubble to the surface and explode.

My mother wanted to be a published author, only she couldn't handle a rejection letter. She'd network, get someone to recommend a publisher, and submit her manuscript. Invariably it would come back with a nice letter praising her writing and saying that they weren't publishing that particular genre at present. They would wish her good luck and she'd put her work away for six months or three years, or until she got another lead.

After marrying my father she had alienated herself from her family. She and my maternal grandmother spent their time arguing. I now understand why. My mother, who was born in Minnesota, claimed to have no birth certificate. She claimed to be born the same year as my father, but thanks to Ancestry.com and their easy-to-access resources, I discovered that she was actually seven years older than she claimed.

This is an important fact for a number of reasons. First, my father had already had another serious relationship with a woman a few years his senior. The reason he didn't marry her was because of their age difference. Enter my mother who was probably even older than she was. If my mother's real age had ever come out, my father would've walked out without even blinking. And not because of the age difference, but because my mother, who couldn't tell the truth, always prided herself on making others *tell the truth.*

To make matters even more intense, my mother chose the birth year of her younger sister, and to protect her secret, she had to alienate herself from her whole family. I was on my own.

Sadly, my father's parents died before I could get to know them. I had an aunt and an uncle and cousins living in and around Los Angeles, but we seldom saw them. My father was a womanizer, which is understandable since I don't think he and my mother ever spent one pleasant day together. The good news was that he chose his employees well and they were loyal. He could go hunting and fishing knowing that his business was in devoted hands.

I was about to start junior high, as we called it then, when Wendell Corey's family moved in around the corner, three houses away. Suddenly my life opened up. I found myself around happy, engaged, inspiring people. Wendell had recently filmed *Loving You* with Elvis Presley, and he used to pop by their house with his "Memphis Mafia" in tow. Since Wendell had starred in movies as well as on Broadway, there were always New Yorkers streaming through the Coreys' parties.

Wendell and his wife, Alice, came to a school play, and afterwards the actor looked at me with his piercing ice blue eyes and said, "Linda, you can do anything you set your mind to." That moment has been the anchor of my survival.

I had the superficial trappings of white privilege, but after I graduated from USC I hit a wall. What was I supposed to do now? I wanted to write for film or television, but I didn't know where to begin. I had very little self-confidence, which led me back to my parents' house in Brentwood. My moral compass was all over the map.

When my father had an affair with one of his employees I realized that he cared more for *the other woman* than he did for my mother. The message I received was loud and clear: If you want a man to love you, be the other woman.

Of course, today I no longer believe that, but as an impressionable, unhappy college student, my parents' dysfunctional relationship and separation had taken precedence over everything else. Even when I escaped to Traverse City, Michigan to be an apprentice in summer stock, my mother wouldn't leave me alone.

One pleasant day in July, I received a letter from her letting me know that *the other woman* was trying to murder her—that the garage door had fallen on her, nearly killing her, and if she should die before I arrived home, I was to have her death investigated as a homicide. My mother never let me out of her emotional grip.

Laser Lady Meets the Light Junkies traces the times and events that have shaped me as an adult. From 1972, when I saw a small hologram of a cannon floating inside of a plastic cylinder, to Salvador Dalí requisitioning the first multiplex hologram of Alice Cooper, and to the individual pioneers of a new three-dimensional medium, I will give you a brief history of holography along with my enthusiastic attempts to find success as a screenwriter in Hollywood. It's a speeding roller coaster without safety bars, so I hope you enjoy the ride.

CHAPTER 1

LIGHT JUNKIE

The sun was rising and the moon was still in Libra when the couple I had just met faced each other, smiling as they locked eyes. It was February 5, 1972 and a small group of new and old friends had gathered on a chilly bluff overlooking the Pacific Ocean in Santa Monica to witness the marriage of two flower children.

We were living the dream. We were well-educated twentysomethings who believed the world could be healed with peace, love, and tolerance. Call us hippies if you like. We looked the part with flower garlands crowning our long hair, long skirts, strings of handmade beads, embroidered Indian blouses, and faded jeans.

The very pregnant bride and her groom had flown out from New York accompanied by a few close friends, and I thought it would be fun to watch my friend Jim, a first-time mail-order certificate minister, perform the nuptials. What I had no way of knowing was that this innocent, random adventure would end up changing my life forever.

It all started at the short breakfast celebration after the ceremony when I was introduced to an energetic New Yorker named Cary Lippman. He had just relocated to L.A. to open a nightclub called the Paradise Ballroom.

"Have you ever been to the Electric Circus in New York?"

"No," I told him. I hadn't even heard of the Electric Circus. I was a

California girl, and based on past experience, if something was wildly successful in New York, it had to be really amazing. L.A. was less about nightclubs and more about listening to performers in small venues. We had the Troubadour on Santa Monica Boulevard, and the Whiskey A Go Go, and the Roxy a couple miles north on Sunset. Industry types: musicians, actors, agents, writers, filmmakers, and a smattering of tourists would crowd into these night spots to network and breathe the air alongside Linda Rondstat, Jackson Browne, or Billy Joel.

These were heady times, pot-smoking times. Vietnam was winding down and the Mothers of Invention were listening to the Grateful Dead.

Surrounded by the handful of wedding guests, Cary described what sounded like L.A.'s next big thing: The Paradise Ballroom. He painted a picture of a dilapidated two-story shopping plaza off Santa Monica Boulevard in West Hollywood that was being transformed into a state-of-the-art bacchanalian playground. It had a huge parking lot. The dance floor would consume the length of the building, while the same bands that had played the Electric Circus in Manhattan—the Grateful Dead, Sly and the Family Stone, the Chambers Brothers, and the Velvet Underground—would perform while hundreds, maybe even thousands, danced until dawn. This was groundbreaking. The more Cary described the new club the more excited I got.

"And we're thinking about having a children's theater on the weekend."

Well, that was all I needed to hear. I'd majored in theater at USC and played Cherry the Clown in summer stock. After college I'd put on puppet shows at children's birthday parties. My friend Jim and I looked at each other.

"I would love to write an original children's play!" I eagerly offered.

"I've got production experience," Jim volunteered enthusiastically.

"Wow," Cary was nodding his head. "Let's talk Monday."

Over the next couple of days I asked people I knew to describe the Electric Circus. They all told me about its mastermind, a former William Morris music agent named Jerry Brandt. He had a reputation for being both brilliant and hot.

The Electric Circus had closed in September 1971, which explained the timing of the West Coast foray. To quote Wikipedia, "The Electric Circus became 'New York's ultimate mixed-media pleasure dome,' and its hallucinogenic light baths enthralled every sector of New York society. Its hedonistic atmosphere also influenced the later rise of disco culture and discotheques." They had light shows, circus performers, experimental theater…my head was spinning.

I needed to have an idea in mind for the children's theater. Something that would appeal to both children and adults. Something with an educational twist. Cary had mentioned wanting to placate the surrounding residential community with family entertainment.

Los Angeles was a cosmopolitan city, so why not create characters who could visit other cities, even countries…? By the end of the weekend, I'd come up with "The House That Went Everywhere." This traveling "house"—a set that we would create in a corner of the venue—could be a backdrop for multiple storylines, plus it would provide elements of diversity. It could be a neighborhood crowd pleaser.

Monday morning Cary called and told me to meet him at the building on La Peer. It was, as he described, a rundown shopping mall with a large parking lot. The ground floor, like some abandoned tourist outpost, still had a few tenants selling t-shirts and candy. The shops on the top floor had been gutted to make room for a massive dance floor.

Cary saw me and waved. I listened to the rasping saws and hammering, and watched the long-haired boys of summer adapt to winter, wielding hammers and yelling at each other above the din. There was a sense of purpose, of urgency. Even the noise had a nice "there's no business-like show business" ring to it. After being told what was going to fill this vast expanse, Cary introduced me to its oracle, Jerry Brandt, a man in his thirties who exuded animal magnetism with the aura of a rock star. His face was angular, sculpted. My heart skipped a lot of beats. And, as I later found out, my friend Joan Nielsen, who signed on as an artist, had the same visceral reaction.

"Hey," a diminutive man with a strong Southern drawl called over the racket. "You're producing the children's theater?"

I nodded.

"Do you like magic?"

"I love magic!"

"Good!" he said extending his hand. "I'm Charlie Patton and if yer gonna produce the children's theater, you gotta use holograms!" He looked at me quizzically. "You ever seen a hologram?"

"No."

"Follow me!"

Charlie Patton looked like an enthusiastic gnome and sounded like a good ol' boy with a residual Tennessee twang. As I soon discovered, he'd been brought to L.A. by Elvis and his Memphis Mafia. He was thirty-three and divorced, with a young son and a reputation as a very talented art director and set designer. Charlie kept his finger on the pulse of music, film, art, and technology. The Troubadour was his local hangout, and tobacco heiress Doris Duke was his late night, go-to phone friend. That first day I followed Charlie up a narrow, newly built wooden stairway to his makeshift studio where sketches, notepads, color palettes, and tools fought for space. He picked up a six-inch cylinder, moved it around a light until he found what he was looking for, then handed it to me. "You gotta let the light hit it just right," he instructed. As I moved the cylinder he said, "A hologram is a lensless laser photograph. 'Holo' means whole and 'gram' means message. It's 'whole message' in Greek."

Suddenly, in the middle of what had been empty air there was a small metal cannon. It appeared to have weight; to be real. I looked at Charlie and he looked back with a twinkle in his eyes and a smile the size of Texas.

Mission accomplished. I had never seen anything more astonishing. My mind started racing through all sorts of 3-D images and applications. I wondered, what if someone could make a larger-than-life-sized image of a person that would appear to float off a movie screen?

Charlie and I became fast friends. His passion for holography was contagious, only he thought in terms of music and the visuals that could be produced in a giant arena or a stadium, while I thought of scenarios that could utilize this new 3-D technology as a movie plot device.

I asked myself, what would be the most astonishing vision imaginable? The answer was simple: the Second Coming of Christ. There were millions of people anticipating that day, and if someone could bring it about using a hologram, the quotient of true believers would rise exponentially. A plot had begun to take shape.

Then one night at Dan Tana's, a popular Hollywood haunt within walking distance of the Paradise Ballroom, Charlie introduced me to a charismatic young artist named Ed Ruscha. At that point Ed was known for pop art and his work titled "Standard Station." Today Ruscha's work hangs in the most important collections and contemporary art museums in the world.

Sitting next to him at Tana's that night, looking into his deep blue eyes and feeling his electromagnetic energy, I was inspired to create a key protagonist for *Revelation I*, only now, I felt like one of my protagonists had jumped off the page, and as if by osmosis, turned imagination into flesh and blood. I believed if moviegoers could experience my cannon in the cylinder epiphany they would demand more three-dimensional experiences, and maybe even 3-D movies without having to wear red and green glasses.

The main character in my script was an artist who finds holography fascinating and confounding. He's not a physicist. He doesn't know how to make a hologram, but he does understand the spatial relationship and impact of three-dimensional objects that appear to be suspended in space, and the surreal sensations they evoke.

He needs a sidekick, a crazy genius who sees the future but can't realize it because he lacks social skills. His behavior frightens investors away. He needs the artist as much as the artist needs him.

As I watched Charlie plead with actors Dean Stockwell and Tom Baker to see the world through his eyes, I realized that he would be my inspiration for the artist's sidekick. Charlie Patton fit the bill. He was on a mission to transform visual space using holography. He was my new science guru and he lived and breathed and drank to every technological advancement. He was my guy.

To me, my screenplay, which I dubbed *Revelation II*, was unfolding with obvious certainty. These two unlikely allies would have to find the money to make a larger-than-life-sized hologram, not to mention buy a pulsed laser and enormous holographic plates. They needed thousands of dollars and a financial visionary with very deep pockets. Enter Father

Fantastic, the charismatic black preacher sitting on a mountain of gold.

If you lived in Los Angeles in the early 1970s, you couldn't miss Reverend Ike, the flamboyant black preacher who drove a Rolls Royce and convinced his parishioners that the more money they tithed, the more money they would receive. Wealth was God's way of rewarding his most pious and deserving followers.

Revelation II was the story of an artist who sells a preacher the second coming—a larger-than-life-sized hologram of Christ—for his Easter morning service inside Houston's Astrodome. Father Fantastic and his Sacred Order of the Spirit (SOS) church would be deemed the direct conduit to God.

There are plot twists leading up to the moment when the image of Christ on the movie screen suddenly and inexplicably floats off and hovers near the audience for up to thirty seconds. No 3-D glasses required. That was my objective, and according to Charlie there were people in American laboratories who could make this never-before-seen special effect happen in first-run theaters.

On February 18, 1972, I registered *Revelation I*, an original synopsis for a screenplay, with the Writers Guild of America, West. I now had two major projects: writing and producing a children's play, "The House that Went Everywhere," for the Paradise Ballroom, and finding a producer who would make a development deal on *Revelation I.*

There was only one little hurdle. I needed to pay my rent and Jerry Brandt was either not around when money was to be discussed, or he was conveniently caught up in the throes of something urgent. Everything, I would soon find out, was being done on the cheap. Carpenters were being paid $75 per week instead of the $200 per week union rate. The backer behind the scenes, behind Jerry, turned out to be Bernie Cornfeld, the controversial international financier who made billions selling mutual funds.

Bernie Cornfeld's financial empire today would remind one of Billions, the Showtime drama that has the federal Securities and Exchange Commission (SEC), chasing the players who skirt the edges of the law. In 1967, IOS, Cornfeld's mutual fund, agreed to sell all of its American operations, and by 1970, the company was faltering.

None of these things deterred Mr. Cornfeld from buying a mansion across the street from Hugh Hefner in Bel Air and installing a dozen or so beautiful young sirens. Bernie was setting the stage for his big Hollywood style comeback. The Paradise Ballroom would be his crowning glory.

I'm sure he imagined long lines snaking around the block to Santa Monica Boulevard and celebrities on speed dial begging for VIP admittance. The burning question was whether or not he had enough cash left to support his lavish lifestyle and at the same time build and open the club. Would it be heaven or would it be hell? We were all on the edge of our seats.

CHAPTER 2

PARADISE LOST

February 21, 1972

This morning it took me 4,000 hours to get ready for the Paradise. Jerry is uptight. I sense that's why he's pressing everyone so hard. He's not too sure about Los Angeles and is anxious and skeptical. My own feelings lie with fear. Fear of being good enough to please him.

Reading this diary entry lays bare my lack of self-confidence and self-esteem. My parents' love/hate relationship had taken up so much of my life that there was little left of me. I was twenty-seven years old and still trying to get my father's approval. He was handsome, powerful, and emotionally inaccessible, so I found men with similar attributes and limitations catnip.

Clearly, I was smitten with Jerry Brandt. In my diary I wrote that he had everyone mesmerized, and at this point, he did. In true Keysey-esque fashion—"You're either on the bus or off the bus"—we were all riders streaming towards this man's over-the-top West Coast nightclub dream.

Because the financial arrangements for the Paradise Ballroom were so fluid, every day was a new crisis, and even though the hammers kept pounding, our overall confidence kept eroding. One day I was on top of the world writing an original play for the children's theater, and the next I was told to "stop" and hang tight.

At the last Ballroom meeting we had been informed that the budget for the children's theater had not yet been approved. I was told to come up with an act—something original—something really "out there" for the night show! Since the Electric Circus had successfully created what could be called a psychedelic circus, I went back to my roots—a Fellini clown, a job I actually had on my resume.

While attending USC I had apprenticed at the Cherry County Playhouse in Traverse City, Michigan. I played Cherry the Clown, the beloved mascot of the children's summer theater. Full-face makeup and costuming offered a sense of complete freedom and I thoroughly embraced it.

My second clown was inspired by Federico Fellini and the artistic, often symbolic use of his circus characters. Foreign features had begun to motivate many American auteurs to break from traditional Hollywood formulas to make what were being called art films.

In 1967, I was cast as a Fellini clown in one such passion project. We shot the film on location in Ojai. I don't recall the title or even the storyline…if there was one. I do however vividly remember being handed an antique parasol and told, "Hold the parasol up in your left hand while swatting imaginary butterflies with a badminton racket in your right hand." I was instructed to do this while taking balletic hops down a rocky hillside. The director looked me in the eye and barked: "You have to do it in one take or you'll be left on the cutting room floor." Needless to say, I did it in one take.

The film business could be brutal, and even though the producers were Hollywood A-listers, their foray in auteur filmmaking never made it to the silver screen.

Having enjoyed playing the two previous clowns, I welcomed my new role at the Paradise Ballroom. A wonderful paisley costume with a stiff, layered white crinoline collar was being made for me. I convinced myself that this, the third time, would be the charm. It was either invent an act for the bawdy night show or go home. My character had to stand out and for that I needed a spectacular prop.

Feeling the pressure, Jerry flew his friend Rick in from New York. Rumor had it that Rick had been instrumental in shaping the success of the Electric Circus. In his West Coast incarnation, he had a perpetually runny nose, fraying nerves, and always wore mirrored aviator sunglasses. Using the crumbs left over from the shoestring budget, he was supposed to hire performers who would provide artistic circus-style acts. Those of us who'd been there from the beginning were informed that we had to audition along with the cattle call. It didn't matter that they'd had a very expensive clown costume made for me. Excite the new majordomo, or be kicked to the curb. By late March, Jerry Brandt barely had one hand on the wheel.

Bernie Cornfeld began popping in to check on the club's progress. He brought people like Al Capp, the famous cartoonist and creator of the L'il Abner comic strip. Then he recruited David Stein, the tall, thin, snappily dressed, hawk-nosed New York pimp, to Los Angeles. Stein's girls were some of the most beautiful, best-dressed females I've ever seen. Rumor had it that these well-educated beauties charged between $2,000 and $2,500 for a date.

The erotic, sometimes lecherous atmosphere guided my thinking. I needed an eye-popping prop, and as luck would have it, I found an armless mannequin in one of the vacant shops. First came white fabric arms followed by a complete makeover.

Her freshly painted eyes cast a piercing stare, and her head was wrapped in a Carmen Miranda headdress. I covered her body with a layer of white netting, then sewed individually wrapped candies front and back. I needed this life-sized art piece to move with me, so a handle was strategically screwed into her back, and her feet were attached to a kidney-shaped wooden board on wheels. On one of her cloth arms rested a wicker basket filled with more wrapped candies. She had become *The Eat Me Girl.*

Rick took one look at me in full makeup and costume pushing the Eat Me Girl and shouted, "You're hired! Money's lousy, but it'll get better!"

And that was it. With the wave of a hand, I had been deemed freaky enough to make the cut. We were all giddy with anticipation. Of course, we wanted our back pay, but we were optimistic, believing that once the club opened, thousands of hip people would beat a path to our door.

> Saturday, April 8, 1972
>
> Everything's falling apart. Joan was fired by Dante because he said there was no more money. Wendy's being paid out of Jerry's pocket. Practically NO acts came to audition. It was very quiet, depressing. Ted and Gus looked like death rewarmed. Lyndon is opening a candy store. Jerry massaged a very tan beauty with a short frizz do. I split for the magic shop where I bought a wilting feather flower and a light bulb that lights up when it touches metal.

The day of reckoning was quickly approaching and the general public was being told that the Paradise Ballroom was going to be the opening of the year…and in some strange, perverse way, it was.

Of course, it had to be a fundraiser for a very worthy cause, something both politicians and Hollywood could rally around. The Elizabeth Fry Center, a halfway house for women parolees, fit the bill. The sponsoring committee included Congressman and Mrs. Alphonzo Bell, the Honorable Edmund G. Brown (Jerry's father and former governor of California), Senator Alan Cranston, and Senator John V. Tunney. These were all 24-carat-gold politicians—serious individuals who would have had heart failure if they'd known what was going on behind the scenes. Hollywood sponsors included Warren Beatty, Diahann Carroll, Ossie Davis, Cary Grant, Shirley MacLaine, E.G. Marshall, George Plimpton, and Nancy Wilson.

On one hand, in spite of the fact that no one was getting paid, the workers were killing themselves working 24/7 to finish the job. Whatever dollars were left were rumored to be greasing the palms of city inspectors. And, the invitations were in the mail.

As the opening approached, anticipation was building on both sides—the Paradise Ballroom people were on the verge of a nervous breakdown, while the general public was expecting what we would think of today as a Cirque du Soleil extravaganza.

> April 20, 1972
>
> 8:30 PM: Panic in the dressing room. The sound is still being worked on. Floors swept, dope smoked, guards dressing in Santa Claus suits, crazies mingling in the dressing room. Crowded. Panic in the air. The club might not open...
>
> 9:30 PM: The sound still isn't working. The great Chip Monck is on the scaffolding, in the sound booth, everywhere he can imagine the problem lies. The Sparks Brothers wondering if they should go on at all since the sound system hasn't been checked.
>
> 10:30 PM: Meeting with Rick and Joyce from the Hog Farm in the pillow room. All listening for our new time slots. "You follow the Oingo Boingo Band and that's after the Roller Queen. Rick has the bullhorn. Panic stricken J.B. enters hysterically demanding the bullhorn. The lights in the penny arcade are out. In fact, all the lights on the left side of the building are out, as are every other store. The whole block goes black.

Sharon Kay Koch, a Society editor, reviewed the opening for the Los Angeles Times. She began by saying, "Conventional society may have stayed away (or come incognito) but the young, curious and the freaky came in droves to the benefit opening of the Paradise Ballroom."

She notes that the invitations stated 9:30 p.m.—however, the first paying guests didn't get in until around 10:30, and lines were still forming at midnight. Guests had been told the dress code was black tie or "bizarre." Bizarre won, with transvestites breaking new ground.

Koch went on to report, "...Most everyone stood in the street for hours before getting in. This included the Ballroom's big-daddy backer, Bernard Cornfeld (who gave a pre-opening cocktail party in his new Southland digs) and other luminaries like Warren Beatty, Dave Garroway, George Hamilton, Jack Nicholson, Lou Adler, and former Governor Edmund G. (Pat) Brown.

"But head guru Jerry Brandt, whose last brainchild was the Electric Circus in New York, took care of that. He imported a vivacious band from South Central Los Angeles to distract the waiting guests with a street concert.

"Chip Monck, the lighting expert for Woodstock, was still in the rafters at 11 pm, plugging in the last fifteen miles of fiber-optic lighting that covered the former Factory's beams, girders and braces."

Seasoned first-nighters were used to rubbing shoulders with electricians and carpenters on their way out, only this time most of the workers threw in the towel and joined the party. While the opening turned out to be what backers considered a publicity success, the benefit didn't benefit anyone. In Sue Cameron's "Coast to Coast" column in the Hollywood Reporter she wrote: "'FIASCO, DISASTER AND BUST,' are just a few words that are on the lips of people who went to the opening of the Paradise Ballroom."

Apparently, the public can read, because only a handful of freaks showed up opening weekend for the dance marathon, another of Rick's brilliant ideas. Ms. Cameron predicted the Ballroom would last six months. It lasted less than two.

CHAPTER 3

LIFE AFTER PARADISE

We've all been teased by the fateful concepts of love at first sight, or of visiting a new place and sensing déjà vu—I've been here before—maybe even in another lifetime. The pragmatic among us may not appreciate the fine art of tuning into the unseen. However, for me, it was the ether that transported me to the future. I began to envision the marriage of science and art, of lasers and light shows, of medicine and three-dimensional imaging. The more I saw, the more committed I became to becoming a proponent of this new three-dimensional medium. In my mind's eye I intuitively knew that holography would become an integral part of our daily lives.

Luckily my new friend, Charlie Patton, was a relentless voice—the drip of a faucet, the car alarm that won't stop screaming. In the lead up to the opening of the Paradise Ballroom, Charlie kept insisting that we needed a hologram. Finally, Jerry Brandt flew physicist and holographic pioneer Lloyd Cross to Los Angeles.

> Saturday, April 15, 1972
>
> Walked in—phone rang—Whamo—Lloyd Cross flown to L.A. by J.B. (Jerry Brandt) to discuss making a hologram for the Ballroom.

Less than one week before the Paradise Ballroom's ill-fated launch, it already had seemed doubtful that the holographer's visit would be eventful. I had my doubts. Rick had been out of control, and there had been a bad taste you couldn't quite brush away. Doom had hung like thick smog. Even then, I didn't really care if the club succeeded or failed. I was a light junkie and Lloyd Cross was going to teach me about lasers and light and holograms. He would make *Revelation* work. It wouldn't be just another caper story; it would educate as well as entertain. It would be a revelation like discovering the religion of light—a three-dimensional image would leap from glass as thin as a windowpane to mesmerize people all over the world.

"Have you seen the holograms at Disneyland?" someone asked Lloyd. He hadn't and he wanted to observe exactly what they had done to create holograms in the Haunted House. We drove out to Anaheim and like all the other tourists were shuttled through the exhibit. Lloyd was baffled the first time through. The miniature talking lady's head in the crystal ball certainly looked like a hologram. It had color and parallax—you could see around it—it had motion....

"See the film being projected?" Lloyd asked me on the second go-round. "The image is being projected from more than one source. Very clever."

The most spectacular effect was the turn-of-the-century banquet with dancers. It was as if you walked into a room filled with people who suddenly turned into ghosts. To create that effect using holograms would have cost millions. No one had figured out how to make holograms move. Lloyd had a theory that utilized motion picture film prior to holographic emulsion, but to date, no one had attempted anything larger than a moving fish. The banquet scene, Lloyd told me, was in his opinion done using a special split-screen onto which the subjects were projected using parabolic mirrors. There were no actual holograms in Disneyland's Haunted Mansion.

Relieved that there wasn't a raving genius working for Disney, Lloyd took me to Cal Arts, Uncle Walt's private dream college in the San Fernando Valley. He wanted me to meet Peter Van Riper, who was teaching holography.

Van Riper, Lloyd, and I went to the small holography laboratory and made an 8-inch by 10-inch transmission hologram of a Tlingit Indian basket shaped like a teapot. Instead of a table, Van Riper used a sandbox on the floor. Sand held the objects in place and served as an interesting background. It was a stable medium for making holograms. The holographic plate was developed much the way photographs are immersed in a chemical bath. I walked out and when I stepped back into the darkened space I saw two teapots. They looked identical.

Peter stopped me. "Which one is the basket and which one is the hologram?"

I stood there looking down. Wondering. Finally, I pointed at the basket on the left. Peter and Lloyd burst out laughing. "Wrong," Lloyd teased. Peter lifted the real basket and I realized that I had identified a laser-illuminated re-creation of the object. The hologram looked more real than the actual object. Another lesson: another encounter with one of the pioneers of holography.

While waiting for the Paradise Ballroom to die a natural death, I threw myself into the search for a holographer capable of engineering a larger-than-life-sized three-dimensional image. McDonnell Douglas had produced the first small commercial holograms, and Dr. Ralph Wuerker at TRW was a leading expert in the new field. The question was, were conventional institutions that relied on government funding and research projects going to be interested in working with an enthusiastic neophyte screenwriter? They had no skin in the Hollywood game, nor did they need to. I had to be on the hunt for rebels—rogue physicists who wanted to think outside of the corporate box.

Lloyd Cross and Jerry Pethick turned out to be those brilliant revolutionists.

My enthusiasm ignited interest from a New York producer named Peter Cookson whom I'd met when he produced *A Streetcar Named Desire* in Los Angeles, starring his wife Beatrice Straight.

I showed him my *Revelation I* treatment and told him what I wanted to do with the screenplay. The concept intrigued him and although it was a gamble, he was willing to take a chance. We made a development deal. I was to get $1,000 for the first six-week option, $1,000 for a second two-month option, and $5,000 after that for a six-month option. In 1972 that was enough money to support me while I turned *Revelation I* into *Revelation II*, a feature-length screenplay.

> Tuesday, April 11, 1972
>
> Left early Tuesday morning with Charlie (Patton) for San Francisco. We took a bus into town from the airport, then a trolley car to Ghirardelli Square for a fantastic Italian lunch with wine, then we found our way to Shotwell St., Lloyd Cross and the new School of Holography. The minute I met Lloyd I was bowled over. Seemed I've known him in another life. Even his partner Jerry (Pethick) struck me in the same way. I was most impressed by their plans for four 4'x5' camera rooms and a pulsed laser.

Lloyd was in his mid-thirties with a compelling, illuminating, toothy smile. His hair was a wispy brown bob. Although of medium height and build, he commanded the space around him. His genius was real and those fortunate enough to be part of his phenomenon knew it.

Jerry Pethick was more circumspect. He was a no-nonsense Canadian artist who rolled his own and was clearly unconcerned about up-to-the-moment dental work. He looked at things through an artist's lens.

From the School of Holography in the Mission District, Charlie and I piled into Lloyd's Keinholtz-type station wagon—one side was totally gutted—and we headed for Jerry's house. The atmosphere was electric. Jerry's lovely English partner, Margaret, made dinner for all of us.

> Tuesday, April 11, 1972
>
> I was flipped out over the dark room downstairs. Pure granular ruby red laser light. The most sensual thing I have ever seen—touch the sandbox and the light pulsated in orgasmic patterns. Lloyd is beautiful. Patient. He explained everything beautifully in a soft, kind voice. Lloyd took my face and gently positioned me in the laser light. He wants to shoot a hologram of me.

I had grown up in a bubble overseen by my suspicious, overly protective mother. She listened to my phone calls and read my diaries until I started writing them in French. Suddenly, when I experienced a room filled with granular red laser light, the veil of self-doubt began to lift, and I felt like I was emerging from a constricting cocoon. Lloyd Cross made me feel both beautiful and smart.

On one hand, I was becoming consumed with lasers and holograms. On the other, I was trying to gain a foothold in Hollywood. In the early 1970s there were very, very few women writing for film or TV. It was an old-guard man's game. And I don't mean a gentleman's game.

A couple years earlier I had become friends with an actress who knew I was interested in writing screenplays and who introduced me to Shelly Davis, a producer friend of hers.

"He started the Whiskey—you know, the Whiskey A Go Go on the Sunset Strip," she recalled. "Now he wants to produce movies and TV and he's looking for writers. Meet with him. See what happens."

I met with Shelly Davis and his partner, Shelly Brodsky, a former William Morris talent agent. They were about to form a production company called Now Productions, and if they liked my writing I could be their in-house screenwriter.

Shelly Davis was a rotund man with a passion for cooking and a gift for marketing and publicity. He was always hustling, always pitching ideas, always onto the next project. He and Shelly Brodsky, who was still managing comic Soupy Sales, made a deal with MGM to develop properties. In exchange for office space, the studio would have first right of refusal on anything Now Productions developed.

In order to get an agent and land a writing assignment, you had to be a member of the Writers Guild. The catch was that you couldn't join the union until you sold something to a company that was a signatory. For me this was a big deal. Joining the Writers Guild of America West not only meant that I had arrived as a professional writer, but it gave me life-changing guarantees—and Now Productions was going to give me my golden ticket.

I aligned myself with the two Shellys and started writing for them. At the same time, my friend Joan Nielsen had gone to work for Karl

Fleming, the former West Coast bureau chief for *Newsweek*.

Fleming, a journalist celebrated for his down-in-the-trenches civil rights coverage, aspired to start a publication with the punch, grit, and gravitas of *Rolling Stone*. He saw it as an opportunity to develop fresh, edgy Los Angeles talent. He enlisted philanthropist Max Palevsky, who had recently sold his computer company for $920 million and had aspirations of becoming a movie mogul. What better way to dip his toes into the film industry than to back *L.A.*, Fleming's new weekly newspaper.

> June 16, 1972
>
> Went to office then out to lunch with Joan and Karl Fleming. He's interested in my diary— wants me to write a column which he will make into a book. I was very uptight, then I decided that I could and should trust him. Went through 3,000 changes trying to rationalize giving him my diary. Nice man, exceptionally handsome, charming. Joan and Blue think paper's disorganized. Skeptical. It's a lot more together than the Ballroom.
>
> June 30, 1972
>
> Go to the office and type more columns. See Karl Fleming. He reads and revels in the execution. The column may be my "execution" someday. I have grave trepidations even though I'll use a pseudonym and will manufacture fantasies. My grave error: He asked, "What's the minimum you can write these for per week?" I didn't say anything. I didn't open my mouth. Then he told me $25 per week and fifty percent of the book.

My column for *L.A.* became a bone of contention. Karl Fleming wanted dirt on the Paradise Ballroom—payoffs, drugs, code violations. When I told him I didn't want to write about those things, he said I was being sophomoric. I ended up writing a few harmless columns and then having to fight to get paid. I finally had to threaten to take him to small claims court.

While this was going on, I was finishing *The Pollution Solution*, a treatment for a weekly TV series that I was developing with the two Shellys. They gave it to Carl Reiner, who said it was even better than described. Reiner sent it to his agent at William Morris, who sent it to 20th Century Fox. With all these irons in the fire, I was sure that something was about to click.

At the same time my quaint little bungalow in Westwood near UCLA was becoming too noisy for my grey-haired neighbors. The units formed a circle of small, single-story apartments that faced a landscaped courtyard. Mine was a studio with a Murphy bed (that, when open, filled the living room), a kitchen and a bathroom. In fairness, the walls were thin and I was leading a 24/7 lifestyle.

Artist and restaurateur Ardison Phillips might show up after closing Ports or the Studio Grill. He and another artist had created a huge infinity installation at Cal Tech. One night Ardison started playing the Staples Singers and drawing on the wall. Another night, Byrds drummer Michael Clark showed up at 3 a.m. It was always something or someone, and I was evicted.

Through a friend I was introduced to a woman who was going through a divorce and wanted to rent the bottom floor of her Woodland Hills home. She was a secretary to one of the heads of a major talent agency.

The house was in the hills—rustic and refreshing, and in those days there wasn't much traffic on Mulholland Drive or Coldwater Canyon. You could hear the birds in the morning and the crickets at night. Aside from the occasional rattlesnake, I thought it would be an ideal place to write. I had found my new home.

CHAPTER 4

HOLOGRAPHY VS. HOLLYWOOD

Let's examine the ingredients in this stew. Certainly there's a good deal of pot smoking, which leads to bouts of wild imagination, both personal as well as work-related. Sprinkle with ambition, stir gently, and you get an adventure that brings scientists to the altar of filmmaking—and filmmakers to the altar of science.

While pursuing the individuals capable of making a larger-than-life-sized hologram, I dealt with a number of scientists who fell into two categories—first and foremost, the "my way or the highway" diehards and second, corporate career professionals who acted within a narrow scope of bureaucratic precision. In any event, it was a learning curve and I knew if I wanted specific results, I had to be both patient and persistent.

> August 6, 1972—San Francisco
>
> Sunday we got up around noon—ate a typically hearty Lloyd (Cross) breakfast—he wrote a shooting script and we went to the school (School Of Holography) and shot film through trip glasses. 3-D infinity glasses—3 points (of light) going out 3 coming in—when light hits glass it creates spectral effects. Rainbows. Went to the beach— gathered Pim—Hollander, Naomi—18, been around, zany,

screwed up—searching, Lilliana, 24, nice, attached to Naomi and a few others. Very exciting. Took piece of mylar— used as a tablecloth in North Beach coffee house—had an absolute high time filming!

Monday—went to school (SOH)—made a hologram. Lloyd and Gary (Adams) made new 100 mic trip glasses plate. Three dudes from PG&E (the electric company) in uniform came over with a bag of grass, got stoned, extolled virtues of holography—want to learn—told us how to set the meter back. Monday night Peter Lee came down and we rehashed S.F. He told me to send The House That Went Everywhere to Golden Gate Park for possible production. Met cinematographer Tim Metzger, he wants to film a documentary on SOH. Went back to Lloyd's stoned on new glasses.

Infinity glasses were simple to manufacture and they gave viewers a visual high that mimicked a kaleidoscope on psychedelics. Anyone, anywhere could put them on and see rainbows emanating from every possible light source. The diffraction grating material was a hologram of a pinpoint of light, so when you held it up to a headlight at night, or bright sun, colors would explode, spectral patterns would pop. If you held it up on a grey, overcast day you might see a few muted spectral patterns.

In daylight, the diffraction gratings responded to the sun glinting off chrome or flames from a barbecue.

Nighttime brought darkness and with it a million light sources: headlights, taillights, streetlights, the moon, a regular light bulb, a flashlight, fireworks—virtually any light source became a mesmerizing light show. Infinity glasses equaled a legal, natural high—they were the obvious avenue to finance dozens of other holographic products, including gift wrap and eye-popping business cards.

Lloyd Cross and his acolytes at the School of Holography were an extraordinary bunch. What they lacked financially they made up for with enthusiasm and ingenuity. Like their two leaders, Lloyd Cross and Jerry Pethick, they were outliers.

Lloyd was born on October 7, 1934, in Flint, Michigan. He graduated from the University of Michigan in 1957 with, as he boasted, an exact 2.0 GPA. Translated, it amounted to average, but to Lloyd, the confident visionary, it was exactly what he needed to move forward.

After graduation Lloyd landed a university research associate position working with C. Kikuchi on the ruby maser (the forerunner of the laser). Next came the development of the ruby laser (the first working laser, made of synthetic ruby crystal), and in 1960 he led a development team that built the world's third ruby laser at the Institute of Science and Technology at the University of Michigan. The first two had been built by Hughes and Bell Labs.

According to Lloyd Cross's biography, which he dictated to me when we were putting together the prospectus for infinity glasses, he had accomplished the following:

- 1961: Built and demonstrated the first high-power ruby lasers having the capability to drill holes in hard material.
- 1961: Formed the first successful commercial company to produce and sell ruby lasers.
- 1962: Developed the first saturable filters for use in fast-pulse ruby lasers that are now used in making holographic portraits of people.
- 1962: Invented a unique optical system for laser machining the laser axion optic.
- 1962: Co-invented the process of laser spectroscopy with Dr. Peter Franken.
- 1963: His company was awarded the IR 100 award for a new product, the laser microprobe, which was based on the laser spectroscopy invention. The product was one of the first successful commercial applications of the laser.
- 1965: Cross developed an air-sea rescue technique using pulsed lasers, but funding was cut off due to the Vietnam War.
- 1967: His interest turned to holography and the visual arts. He started an art gallery-studio to work on his own.

The real turning point came in 1969 when Lloyd teamed up with Canadian artist/sculptor/holographer Jerry Pethick to organize the first major exhibition of art holograms at the Cranbrook Academy of Art in Michigan. It was successful, motivating the duo to venture into the heart of the art world, New York City. In 1970, they produced "N Dimensional Space" at the Finch Museum in Manhattan.

Imagine you are following paperwork that's over forty years old, and you decide to Google the individuals it's referencing. What's so astonishing is from this very small, very immediate act, suddenly you have a window leading from the past to the present. You find out how important Cross and Pethick are today. You discover that Jerry Pethick isn't just a notable Canadian contemporary artist and sculptor—he's a national treasure.

That news stopped me in my tracks. I remember the first time I met Jerry and Margaret at their rented house in San Francisco. He was a hard drinking, roll-your-own visionary with stained, uneven teeth. He and Lloyd had created a holography studio in Pethick's basement—a sandbox on tires. Using a ruby laser and being able to place objects on what had been conceived as a vibration-free sand table, the Chelsea College of Art-trained painter was able to make holograms.

One of the nicest and most talented holographic artists I have had the good fortune to meet is Ana Maria Nicholson. Her husband, Peter Nicholson, was a sculptor whom she says, "...thought holography with its ambiguity of space and volume would be useful for his work."

In memory of Lloyd, who passed away in 2012, Nicholson has said, "Lloyd Cross, the genius behind holography was in our building! (in New York City). He had been a physicist at the University of Michigan and been involved in the development of the masers and lasers which led to the rediscovery of holography. He dropped a lot of acid and decided that he was finished as an academician and that holography had to be taken out of the laboratory and brought to the general public."

Jerry Pethick and Lloyd Cross were only two years apart in age, both were fathers with small children, and both found themselves struggling to survive while turning their visions into something tangible.

In 1972 the School of Holography was supposed to give them a stable base from which to invent and create while generating income, but it didn't take long for the freedom of the times to overshadow economic stability. In 1975, Jerry Pethick went back to Canada, to Hornby Island near Vancouver. Lloyd stayed in San Francisco.

For him there was something intoxicating about the West Coast. It might have been perpetual sunshine, then again, it could have been the creative hippie environment. Whatever the reason, San Francisco was at the forefront of innovation, be it music, or as in this case, holography.

Lloyd was the captain of the ship. Everyone knew they were at the right place at the right time—they were part of a history-making endeavor. Even before moving to Shotwell Street, while he was in New Mexico, Lloyd made his first integral holography test using thirty photographs of a young woman. The results convinced him that he could build a machine that would turn motion picture film into a three-dimensional, floating, moving image.

I can remember him looking up as if communing with some special, invisible collaborator, and blinking. He blinked a lot. He also smiled a lot. He was a serious physicist who was morphing into a visual artist.

We thought infinity glasses were going to make us all rich. All we had to do was start manufacturing and marketing them and we'd generate enough revenue to keep a roof over the School of Holography as well as allow Lloyd to build his inventions.

That was the plan. I would go home to raise money and come up with marketing strategies.

When I arrived back in L.A. that Tuesday night I was manic. Who could possibly resist diffraction grating glasses that transformed every light source into a spectral array? There was no need to smoke a joint, all you had to do was put the glasses on.

August 10, 1972

Critical day—called Lloyd Cross, got all worked up over School of Holography documentary. (Bob) Gilbert came over about 6:30 and stayed til after 11. He said that he wants to produce Revelation II. Says I shouldn't make a documentary on school cause I should spend all energy on producing Revelation. I'm in a strange place. Don't know which way to go.

I went to University High School with Bob Gilbert and we stayed in touch. In 1970 he won an Oscar for *Magic Machines*, a documentary he wrote and narrated about his incredible kinetic sculptures. The film also won the Prix Corte de Film at the Cannes Film Festival, and was subsequently packaged with *Easy Rider*.

Peter Cookson had given me the breathing room to complete the first draft of *Revelation II*, only now he was completely focused on a new play he was trying out in Boston. I needed someone based in L.A. and Bob Gilbert seemed to be that person.

There was no reason why we couldn't make two films: *Revelation II*, a feature, and a documentary on the School of Holography. I went to work on an AFI (American Film Institute) grant proposal.

Saturday, August 26, 1972

Lloyd flew into L.A. about 8 PM. I picked him up, went home, got stoned, ate dinner. Lloyd didn't feel too shiny. Sunday morning he was violently ill. So ill with flu that by late afternoon he was still too sick to go to the Leon Russell party and concert.

The Leon Russell concert was a huge event. Russell had written a number of hit songs: "Delta Lady," "A Song for You," and "This Masquerade." The musicians on his first album included Eric Clapton, Ringo Starr and George Harrison. I think most people remember him for organizing and performing with Joe Cocker on the Mad Dogs and Englishmen tour. He had established himself as an iconic figure in the music community.

In 1972, Leon was every inch a rock star with his shoulder length dark silvery hair, beard, mustache, and soulful, penetrating eyes, often hidden behind shades. His voice echoed his Oklahoma roots. He was perfect for my pivotal character, Father Fantastic.

Sunday—Leon Russell concert incredible! Jack (Calmes) called and said to be at the Chateau Marmont at 4:30. Lloyd was too sick to go. I got to the Chateau stoned. Too stoned. Walked in and saw Marjoe (child prodigy preacher)—freaked, started shaking. I'd had a flash on the way over that they'd be shooting a documentary (on Marjoe)—they were.

After the concert went backstage, found Jack who adamantly said, "I want to produce your film—use Leon as Father Fantastic and Marjoe as the artist." Saw Marjoe and told him to put the trip glasses on—again we were in the wrong light.

Leon Russell's party taught me a life lesson: Getting stoned was a mistake. Having to go to the party without Lloyd made me insecure and getting stoned made my insecurity worse. Note to self: No more chemical interference.

On Monday Lloyd was still sick. I told him about Marjoe, who was the hottest personality in Hollywood. As a child evangelical prodigy, his parents capitalized on his unique acting ability to make themselves rich. Realizing that he'd been used by them, he made a documentary to expose the unsavory practices and tricks of the revivalist trade. A few months earlier, *Marjoe*, the documentary, had won an Oscar.

Acting in movies seemed like a natural next step. I told Lloyd I'd shown him the diffraction grating glasses at the Chateau Marmont; however, in monochromatic light, they didn't do much, leaving him unimpressed. My second attempt backstage was also in ambient light, yielding little more than a few stray rainbows.

Tuesday—Met with Jack Calmes and Lloyd (Cross) at the Continental Hyatt House for breakfast. Discussed possibilities of Jack using Lloyd in Dallas for Bloodrock Concert effects. Jack said Denny Cordell's into my script—wants Leon to star—could work deal with Gilbert. Took Lloyd to the airport. Went home and worked on School of Holography documentary grant proposal.

Exuding an air of confidence mixed with exhaustion, I delivered the

AFI grant proposal to Greystone, the old Doheny estate in Beverly Hills. An AFI employee and I chatted for a few minutes, and in that short time I got the feeling that she thought my project was so viable that I could find financing elsewhere. The exchange left me cautiously optimistic.

September turned into October and things were moving sideways. Bob Gilbert was anxious to make a deal. He wanted Jack Calmes involved because Jack was a shrewd Dallas businessman whose company, Showco, handled all the sound and lighting for everyone from Led Zeppelin to David Bowie to Leon Russell. Jack and Denny Cordell, Leon Russell's manager, were tight. This would wrap everything up into a sweet package. Leon could write the music plus play Father Fantastic.

I was riding high on hope.

In the meantime, I'd been meeting with agents, seeing movies nearly every day, and making a deal with Lloyd Cross to manufacture infinity glasses.

> Thursday, October 5, 1972
>
> Waited all day for phone calls. Gilbert called early and I called Dallas. Bob said, "I'm quoting: My director says that your script is the best script he's ever read." I wait for Jack to call back. Charlie invites me to a screening. No! I have to wait for a phone call.

Clearly, there was a lot going on. The School of Holography documentary script opened with thirty-seven-year-old Lloyd Cross standing over a laser-illuminated sandbox with four new students breathlessly looking on. He puts an 11-inch by 14-inch plate into a holder and the film's title is displayed in 3-D. A narrator begins to describe the process when one of the students says, "It's magic."

In his slow, soothing voice, Lloyd says, "Well, you know, it's funny. A hologram appears to be magic because it's suspended in space. But, as you'll see today, holography is a simple, calculable process. They're no mysteries! Light is a pure wave. It travels at a constant speed so precise equations leave nothing unexplained."

Over a montage of holograms and laser patterns the narrator says,

"Dennis Gabor made the first hologram in 1948 using a complicated white light source. It wasn't until 1960, twelve years later, that the laser was invented, providing a coherent light source necessary for making holograms."

The proposed documentary mixed visuals with historical content and future projections. It showcased the "multi-plexer," the predecessor of the multiplex camera that would turn motion picture film into a 3-D movie. The narrator says, "The ultimate use of holography will be holographic movies. The filmmaker will be able to create an entire world out of light. The audience will be able to see the three-dimensional creation without special glasses or gimmicks. Holographic movies and television will eventually revolutionize the entertainment industry."

Seeing the future as clearly as I did was a gift and a curse. Intuitively I knew what the future looked like, but I wasn't a physicist and I couldn't physically make what I envisioned. I could only describe it using words.

My documentary script called for a visual depiction of how to shoot a hologram, develop it, and project it. It laid out future applications that included teaching students using three-dimensional imagery, storing the entire Encyclopedia Britannica on one 8-inch by 10-inch holographic plate and setting up a pulsed laser portrait studio to capture the real life 3-D images of human beings. Many of the things we take for granted today were no more than pipe dreams in 1972.

Bringing people together from across the country to make a business deal before the advent of cell phones and computers meant communicating through answering services, rotary phones, and sometimes secretaries. It was a painful waiting game, and I wasn't a terrifically patient person.

Neither was Charlie Patton, my original holography mentor, who was hitting the bottle way too hard. He was committed to connecting people with power and money, such as Doris Duke to artists who were pioneering this incredible uncharted three-dimensional medium. Doris Duke, philanthropist and art collector, Dean Stockwell, movie star actor and friend—these were Charlie's people. He insisted that I call Dean Stockwell. Charlie thought Dean should star in *Revelation II.*

November 1, 1972

Finishing the ending of Revelation—came home and called Dean Stockwell. Smooth voice. He said he'd call me at the office. I briefly explained Bob Gilbert's interest in producing Revelation II.

Dean Stockwell was a child star who had grown into one of the best, most recognizable young actors in Hollywood. His credits included everything from Eugene O'Neill's 1962 *Long Day's Journey into Night* to early television dramas such as *Playhouse 90* and *Alfred Hitchcock Presents*. He was a good friend of Charlie's and a perfect choice to play Jay, the artist in *Revelation II*.

November 3, 1972

Met Dean Stockwell in Topanga. He's so much smaller than I imagined but he's perfect for Jay. We met at a café. I was early—felt like a school girl in plaid skirt amidst jeans and funk. Started talking to a couple poets, then Dean—English cap, rainbow stripes on motorcycle, silver eagle bracelet, Buffalo nickel ring, cowboy boots. We talked about Charlie, the script, got good vibes.

The following Monday I tap-danced to *Gypsy* for forty-five minutes while waiting for the mailman. I'd been told that the AFI verdict would arrive that day, and when it didn't, I called the institute.

"What's your last name?"

"Lane."

"Oh, no," a dispassionate male voice chirped. "You didn't get a grant."

The news was devastating then, and even more so now, because both Lloyd Cross and Jerry Pethick are gone and that important piece of holographic history has been lost forever.

November 12, 1972

Lynn and Jack left for India. Leon (Russell) will read Revelation. If he can get behind it and I have good vibes—we'll spend a week together and write.

November 22, 1972

Go to the office and Gilbert calls. "I have some bad news for you—really bad news...Someone's not your friend. Dean Stockwell took your (Revelation II) script to Tom Cooney at Producer's Studio and told him it was his property." And that Stockwell wants to direct it.

2:30 am Dean Stockwell calls. He's at a bar. He's sorry to wake me. He tells me he took his script to Tom Cooney. Not mine. I hope so!

On Friday, the 24th, I called Tom Cooney, the head of Producer's Studio. One secretary passed me to another secretary who took a message regarding "the Dean Stockwell problem" and said Cooney would call me back. I was on pins and needles.

I wait. I work. Cooney calls, "No, there's no mistake. It's your script!" I drive to Producer's Studio and pick it up. The script is in an old envelope addressed to Dean by his agent. There's no mistake! Dean had pasted his name in my script. Cooney said that he'd like to shoot *Revelation* next. I'm dumbfounded.

On December 24, I spoke to Leon Russell, who was staying at the Chateau Marmont. He hadn't read *Revelation* and he didn't have a copy with him. I dropped off another copy and prayed for a Christmas miracle.

CHAPTER 5

THE BUSINESS OF BUSINESS

It's a new year and I'm optimistic. I call Leon Russell and he says, "I like your script. It's good."

"Well, then, does that mean you're interested in playing Father Fantastic?"

There's a thoughtful pause followed by "...Well, I don't know. The role is so close to my concept of myself I don't know if I want to do it. I would really like to do a straight role first. It's a heavy role, a heavy responsibility. It'll influence a lot of people."

And so it went. I tried to get Leon to meet with me to discuss possible changes—to find a middle-ground solution that would suit both of us. Nothing worked until I hit upon holography. He'd never seen a hologram and wanted to take a look.

> Saturday, January 20, 1973
>
> It's now D-Day. At exactly 3:30, I meet Leon Russell, his black chick, and writer/filmmaker Tony Foutz at the Chateau Marmont. I climb into my ailing Toronado and Leon and his entourage climb into his new Rolls. They follow me to Pasadena where we get lost. Ardison's directions are lousy. I call Caltech—we're around the corner. I climb into the Rolls and apologize for the runaround.

> Everyone's congenial.
>
> We meet physicist Elsa Garmire. Tall. Leggy with a pretty face. Very professional. Good with her two little girls. Talks with her hands. We saw nothing impressive because the lasers were wearing out from too many hours of use. There were only three holograms, my Indian basket being the most effective. Saw the cannon in the cylinder, but overall it was disappointing.
>
> Leon et al. drove off to the Roller Derby. I was disappointed because things were so anti-climatic. We left on up terms. I now must arrange a showing at Hughes or TRW.

Elsa Garmire was a Senior Research Fellow in electrical engineering and applied physics at Caltech when I met her. Her career as a pioneer of laser technology and nonlinear optics, to name a few of her achievements, has made her a legend in more than one discipline. More recently she had spent twenty-one years as the Sydney E. Junkins 1887 Professor of Engineering at Thayer School of Engineering at Dartmouth. Having her introduce Leon to holography was a gift.

I called Elsa to ask if she remembered our meeting with Leon Russell at Caltech. She did. She explained that at that moment in time she was enjoying meeting celebrities. She appreciated the spectacularly beautiful effects of laser light and wanted to share the experience with others.

By the end of 1973, Elsa Garmire, Ivan Dryer, and Dale Pelton had formed Laser Images, Inc. and had opened their first planetarium laser light show at Los Angeles' Griffith Observatory. It was the remarkable beginning of a new art form, and Ivan Dryer had stepped up to become the creative innovator and entrepreneur of laser shows at planetariums and rock concerts worldwide.

Everyone was scrambling. People were coming out of the woodwork with similar projects and running to patent attorneys. I was both excited and anxious. Every day I was seeing something new and exhilarating, and there was never enough time to accomplish what needed to be done. I'd finish one thing only to find six others that needed to be addressed. Sleep? What was that?

> January 23, 1973
>
> Jack Calmes, president of SHOWCO, the company that transports equipment for the biggest acts in rock, is in fact the person who introduced me to Leon Russell. It's now Monday, and I'm working as a temp in the Rock Department at IFA (International Famous Agency). I speak to Jack who tells me he's been to an (East) Indian astrologer and this isn't the year he and Leon will be working together on a film. He tells me to send him the contracts so that he can negotiate with Leon directly.
>
> Thursday, January 25, 1973
>
> There's a screening of Death of the Red Planet at Todd AO, and since I haven't heard from Lloyd Cross or Ardison (Phillips), I go on my own. I start talking to a woman who's the head of children's programming at ABC—I turn around and—there's Leon. He says hello. After a few minutes I make my way over to him and say, "I spoke to Jack and he'll be out here next week to discuss an option/rewrite deal." I tell him the Indian astrologer nixed the picture plans for this year. Leon makes a positive reference and laughs. He was with smiling filmmaker Tony Foutz and dates. Lloyd walked in with Peter Van Riper.

The film ends and everyone looks at his neighbor for a reaction. Lloyd shows Leon holograms. Leon's interested! All happy. Good vibes. I suggest we go to the Studio Grill for a drink. We do. We overwhelm the place. Everyone is jockeying for position. Being the writer of the screenplay I want it to be produced. I have the powerful Leon Russell group on one side and the Bob Gilbert—winner of an Oscar for his documentary—on the other.

This was a particularly intense time for me. The house I was living in had been sold, and moving day was breathing down my neck. I was doing temp work at a talent agency, and I had no idea when or if I would actually receive option money.

That said, there were bright spots. I enjoyed going to Ardison Phillips studio and watching his paintings evolve. The protagonist in my script was a successful contemporary artist, and I needed paintings that reflected his energy and drama. I'd even made a point of introducing Ardison to Henry Seldis, the art critic at the *Los Angeles Times.*

Originally from upstate New York, Ardison and a man named John Platis had created an infinity sculpture that was on exhibit at Caltech. One corridor was illuminated with granular green laser light while the other was granular red. The two corridors met at the Plexiglass apex of a pyramid that created the illusion of infinite space. That is how I remember the piece. It was very impressive and inspiring. Elsa Garmire's expertise had been instrumental in helping the artists accomplish their vision.

> Saturday, January 27, 1973
>
> Waited until three for Lloyd and crew. Gilbert called, "Alexa left me!" He was in bad shape. Difficult to understand as not a month ago she said, "Robert is the best old man in the world." She claims he's too intense and they have a personality conflict.
>
> Took Lloyd Cross group to Ardison's studio to see his paintings. From there we went to Anait Stevens' studio on La Cienega. Gilbert came to Anait's and we all watched Lloyd make a hologram. He's so patient, easy-going.

Anait Stevens, as it turns out, was a force of nature. The artist was a fifty-year-old redhead who, to me, was a very bright, wealthy older woman on a mission to make her own holograms. And she was lucky enough to be married to a tolerant, adoring husband who encouraged her to pursue her fascination for this curious and expensive new three-dimensional medium.

One night a group of us, most of us neophytes soaking up information at the feet of the masters, convened at an Italian restaurant called Bruno's. Anait positioned herself at the head of the table, letting everyone know she was a force to be reckoned with. She loved to flirt with young men, and now, as an older woman, I can appreciate her zest for youthful adoration.

Exhausted, I got laryngitis and lost my voice during dinner. The bill for the twelve of us came to $40. Anait put a $20 down. She knew how to show the physicists that she was not only fun and flirty, but a woman of means. By paying half the bill she was sending a message to the wallet bound physicists: I'm generous, but not foolish.

Eventually Anait, who held a patent for a battery-operated hand mixer, became proficient at making holograms. She quite literally let no man stand in her way.

When she wanted to rent a pulsed laser for $500 a day and was turned down, she bought one for $54,000 and started making extraordinary works of art. What she began in Los Angeles and Santa Barbara eventually expanded to an artist's loft in SoHo.

February 1973—A Shot in the Dark

It was during this window of time that I visited the School of Holography and discovered that the bottom had fallen out. It was winter and a group of us were rattling around the chilly, candlelit converted warehouse. There was no electricity, and the landlord had already started the eviction process. The telephone had a mind of its own. You could ring in occasionally, but under no circumstances could you dial out. The situation was dire and we knew it.

Without electricity we couldn't manufacture anything, and if we couldn't manufacture anything we couldn't fill orders. The only thing that could save this dedicated group of light junkies was an infusion of cash, and so far no one with deep pockets and long range vision had stepped forward.

Suddenly the phone started ringing. It was 10:30 and huddled in candlelit darkness we looked at each other wondering who'd managed to get through. Lloyd answered and listened. He kept blinking and nodding his head affirmatively, until, in his soft-spoken way he said, "Yeah…sure, it's possible. I designed the whole system in my mind… when I was in New Mexico. But I need five hundred dollars to turn the power back on. I can't do anything without power." When Lloyd hung up his face was a beacon of joy and relief.

Salvador Dalí wanted to make a hologram of Alice Cooper from movie film. The 16 mm film would be shot in New York and processed in San Francisco. If his invention worked, if a multiplex laser camera, the Mark II printer, could turn ordinary motion picture film into a three-dimensional, floating, moving image, Lloyd and his rainbow's end associates would have pioneered a viable commercial venture. Jerry Pethick wouldn't have to give up his vision of holographic art. Holographer Lon Moore, one of the brightest and most centered holographers at the SOH, wouldn't have to find another way to support himself while experimenting—teaching himself to make, not just holograms, but good holograms.

The voices on the other end of the line were Cecile Ruchin and Selwyn Lissack, New York visionaries who had sold Salvador Dalí, Alice Cooper, and Joe Greenberg, Cooper's manager, on the concept of a holographic work of art—something that had never before been seen or attempted. The visual future.

To put things in perspective, Selwyn and Cecile had few options—no one had ever converted motion picture film into an integral 3-D movie— and if Lloyd's theory worked, which was always a question, they would position themselves and their company, HCCA (Holographic Communications Corporation of America), at the forefront of advertising, art, and technological advancements. They had thirty days to shoot the Alice Cooper footage and forty-five days to turn it into a hologram that would be unveiled at the Knoedler Gallery in Manhattan.

Few people, if any, could make an integral hologram in this short timeframe. The task entailed shooting the film in a specifically calibrated circular format and then shooting a holographic film from the original footage. In order to see the 3-D result, the image had to be reconstructed using a laser whose beam had been spread through a lens and positioned at an exact angle. Only a rogue genius physicist such as Lloyd Cross would be up for this challenge.

> January 29, 1973
>
> Spent last two days in bed with laryngitis. Gilbert called to say that he spent three and a half hours with Lorimer at Loeb and Loeb. I will receive $25,000 plus rewrite money, plus 5 percent of producer's gross. Better than what we discussed Saturday night.

> Thursday, February 1, 1973
>
> Came home. Made contact with Lynn (Lenau). Would meet at Tana's at 10. Photographer Andee Cohen came with Jack (Calmes) and Lynn. I tried to show them the Infinity lens. We had dinner when Jack explained that Leon would want to produce Revelation with Denny (Cordell) leaving Bob Gilbert behind. I said that I'm inclined to stand with Bob because of circumstances. Fantastic night.

It was decision time. Loeb and Loeb was one of the most prominent Beverly Hills law firms and Walter Lorimer was a senior partner. It was clear that Bob Gilbert wanted to produce *Revelation II* and he was ready to move forward. Jack Calmes and Denny Cordell were a very powerful duo, but would Leon Russell ever be ready to play Father Fantastic? Leon was the key. I had to weigh everything.

Bob Gilbert put together a production team with his friend Don Richetta as the producer. Richetta had produced twenty-five episodes of a TV series called *My World and Welcome to It*. Producing *Revelation II* would mark his moment to move up the ranks as a full-fledged movie producer.

Speaking out, Richetta said there was no reason to move forward until Gilbert had a signed option agreement. I was asked how much I wanted and I said $1,000. Apparently, Gilbert's pockets weren't as deep as he'd led me to believe. He asked if I would take the $1,000 in four increments of $250 per month for four months. Since we were long-time friends, I said yes.

> Wednesday, February 7, 1973
>
> Most incredible day. Went to Loeb and Loeb and signed option agreement. Bob didn't have a check with him so he gave me $40 and said he'd mail me a check for the remainder today.

Needless to say, I read this diary note and was horrified. What was I thinking?

Sure, I was impressed with Gilbert's powerful lawyer, Walter Lorimer, and the fact that I had a signed deal, but I should have taken a closer look at the dark cloud on the other side of the friendly handshake.

To celebrate I visited Ardison at his studio. We'd been inspiring each other for the past six months, and suddenly I thought I was in love with him. All was going well until John Platis, his studio mate and close friend, started talking about "Susan." Who was Susan?

Well, it appeared that Ardison lived with Susan when he wasn't at his studio painting. This was both shocking and disappointing. I asked him if he was in love with her. Sheepishly he said, "Well…I guess I am, but you never know what might happen."

Without even trying I had become The Other Woman. Living at a time when free love prevailed, it was easy to move from one bed to another. Keeping track of one's heart and one's head was more challenging. This was not my first time being swept into the arms of an unavailable amorist. Emotional turmoil embraced me, and I wrote one of my favorite poems.

The Other Woman

The other woman
Chic libertine dancing through chiffon shadows
Into stolen arms

Reaping luxurious benefits:

Charming smiles, bejeweled dials
All the things a girl requires
To live inside a silk-lined drawer.

Now Bob Gilbert and I had something to commiserate about. He'd been rendered an emotional basket case when his wife Alexa left, and I felt confused and betrayed by Ardison.

We both needed to keep our focus on work. I suggested he go to San Francisco and learn how to make a hologram. After all, it was a man's world, and he and Don Richetta were going to be the ones to convince studio chiefs and investors that the very special 3-D effect for *Revelation II* was worth the gamble.

No one had ever made a larger-than-life-sized hologram of a human being that would suddenly appear in front of a movie screen for up to 30 seconds. No special glasses needed. Holography was not the only solution, but if it worked it would revolutionize the film industry.

Walter Lorimer promised that the partnership papers for The Revelation Company, which would be set up to produce my screenplay, would be ready in three weeks. There would be four partners, including Robert Curtis, the director.

I was busy writing an outline for revising the script while Bob toiled with Lloyd at the School of Holography in San Francisco. The Revelation Company would commission Lloyd to make a holographic prototype of Christ that we could use to court investors. That was the plan.

"Wow, Linda! You've got to see what Lloyd's doing—white light holograms! [a new way to reproduce images of 3-D objects]" White light holograms were a colossal breakthrough. Suddenly 3-D imagery could be printed on flimsy film, wedged between two sheets of plastic and viewed in ordinary light.

After a couple weeks in San Francisco, Bob Gilbert had turned into a light junkie and he wanted to be in the hologram manufacturing business.

Up until then holograms had to be both shot as well as illuminated by a laser. It was possible to use a mercury arc light, but they were cumbersome and expensive.

A white light hologram could be viewed by positioning it in lamplight or hitting it with a spotlight. Freedom!

Bob was hooked, and upon his return to L.A., he decided to form a separate company with Don Richetta, a money man, and Lloyd Cross, to mass produce white light holograms. I was not included in this new venture.

To make matters worse, Walter Lorimer structured the new film company—made up of Bob Gilbert, Robert Curtis, Don Richetta and myself—30/30/30/10 percent. Guess who was assigned the 10 percent?

The lawyer's rationale was that the others were going to have to give up part of their percentages in order to make a deal with a major studio and/or find outside financing. At this critical moment I didn't have an agent or a lawyer to negotiate for me. Being a member of the Writers Guild made no difference. I was on my own.

Then on March 6, 1973, the Writers Guild went on strike, and I wasn't allowed to work on my screenplay until it was officially over. My "partners" asked me to bend the rules and I refused. What's the point of a union if you don't stand together and take it seriously? Instead, I walked the Paramount picket line with two writers from the comedy show *Laugh In* and enjoyed the exercise.

Meanwhile, Don Richetta thought the diffraction grating hologram concept— infinity glasses—was well worth exploiting. He arranged a meeting with two potential investors: Terry Andrews, a stockbroker and investment consultant, and Dave Commons, an attorney. Both wanted to be in the film business.

Gilbert, Richetta, Andrews, Commons, and I met for lunch to discuss the potential Lloyd Cross mass production of holograms. Bob Gilbert was now interested in promoting those white light holograms. There were now two companies—Optical Infinities and Fidallgo—but only one Lloyd Cross.

Andrews and Common noted that Lloyd had a whimsical, though genius track record. They told me he was hired by Universal to make a hologram, then went MIA.

When found and pressured, he finally completed the job. What happens, they asked, to prevent Lloyd from taking their investment money then dropping the project for something he invents that he finds more fascinating?

I couldn't answer that.

My emotions were raw. I was up against a group of men who wanted to sideline me so they could claim a bigger piece of the pie. I had to act decisively.

We decided that Lloyd would be responsible for technical oversight, building the equipment and manufacturing the products. I would raise startup money and take care of public relations and marketing while developing distribution channels.

Bob Gilbert and his new cronies would have to come up with their own plan. For the time being I was one full step ahead of them.

CHAPTER 6

PATENT PENDING

Next on the agenda was seeing a lawyer who could set up the Optical Infinities Corporation, and a patent attorney who could start the patent process rolling. The corporate attorney sent me to a Mr. Bronstein, who looked at the holographic diffraction grating in the glasses as well as the glasses themselves and flatly said, "A patent in your case is relatively worthless, a search unnecessary." He said we must nevertheless form the company and then start the patent process, more to evoke fear of patent pending rather than worry about getting the actual patent. (Once you apply for a patent it puts everyone considering a similar venture on notice. If they think you got there first, and are protected legally, they will likely turn to a different venture.)

Monday, February 26, 1973—San Francisco

Went to Jerry Pethick's for breakfast and the pow wow formation of the Optical Infinities Company. Exhilarating! Warm, stoney day. Now I'm really antsy cause I have to get back (to LA) so I can move. I'm uptight about the "scabies" infestation, getting the $37 cash together so we can get a legal partnership.

I'm in an exhilarated state. We talk about raising $45,000 for the company and I internally panic. Then we come back to $35,000. Lloyd and I stand in the chill night waiting for a bus. We're on the wrong side of the street. We finally make it downtown.

"Would you like to get a hamburger?" he asks. We stop at Zim's and I call my friend Susan (Turnbull). She'll meet us there. Lloyd and I discuss success—an Academy Award for special effects for Revelation, the corporate plane. Everything! We have a third cup of coffee, pay the bill and Susan arrives. She's reticent. She drives us to the school, stays 15 minutes and splits. Lloyd and I finish the agreement. I go to bed with Stones blasting. I go to the rear, peer through a curtain at two stoned dudes, one looking like Prince Charming. He jumps up, runs to video room and says, "I didn't want to hear that cut anyway." He turns it off. He is fantastic. I'm enthralled, but I go to bed.

I'm lying there awake, naturally wired. I have to be up and out by 8. Pam (Brazier) walks across to the bathroom. Then, from the darkness the tall Prince Charming emerges. He looks in my direction. I giggle and say, "Oh, I can't sleep anyway". He continues walking across the floor. Sits on my bed—looks down. Says hello. I say hello. He kisses me and it is Prince Charming. We look at one another. Wow. Door opens—Pam and Lloyd nude walk across to take a bath. They look in our direction and smile.

"What's your name?"

"Laser Lady. What's yours?"

"Stonewall."

Beautiful name...long sensitive fingers...large brown eyes. "Oh," he says, "I'd better call Sharon (McCormack) and tell her not to come home."

"What?!?"

He tells me that Sharon's having an affair with a successful San Francisco artist who's a mutual friend. He goes to call her and I wait and wait and wait. It's getting later and later. I walk to his space and ask what he's doing. Brushing his teeth. He invites me in and I enter a quaint, intoxicating split-level abode—parachute suspended from the ceiling, fine antique coffee table. We get stoned, talk. I'm uptight because he couldn't reach Sharon and she may come home. We go to my area and huddle and cuddle, and make love.

I hear Sharon come home. I don't move. She makes so much noise in the bathroom Stonewall wakes up. "I better go talk to Sharon," he says. "I'll be right back."

Half an hour later he reappears. "Sharon thought I should say goodnight to you." Then he continues, "I know it's late and you have to get up at eight, but I wish you could talk to her to make her understand how special this was...I wish you could communicate with her." He thinks a minute then asks, "Would you write her a note?"

"A note?" I've done a lot of strange things, but this... "Okay, I'll write her a note."

We say goodbye and I write: "Dear Sharon, The Little Prince said it best: 'It is only with the heart that one can see rightly. What is essential is invisible to the eye.' As I see it you're a very lucky lady. Sincerely, Laser Lady."

The School of Holography, as you can see from my diary note, was a unique combination of hardworking, creatively curious individuals living in the moment. Youth was not wasted on us. Carpe diem was our motto and each one of us seized each day with gusto.

The cavernous rectangular building on Shotwell Street in the rundown Mission District had been divided up into living and work-spaces. Sharon McCormack, a brainy beauty who was one of the pioneers of Multiplex holography, lived at one end, Lloyd at the other. There were two bathrooms, one with an actual bathtub.

There was an office, holography studios, and darkrooms. Other living spaces were carved out and personalized as needed. The loft belonged to Allen. Mattress, army blankets, old velvet throw pillows. Mylar. Visions of fine art and science mixed with a disregard for order. The dreams, the work, the visions being far greater than the task of ordering a perfect house.

I'd been making frequent trips to San Francisco to document the progress of the multiplex machine. Being a party to what I considered a history-making breakthrough—the very breakthrough that would be used to make a prototype for my film—evoked a continuous sense of exhilaration that I'd never before experienced. I'd been through a lot of manic moments, but this was different. Everything was on the line. If they succeeded, if the machine worked, we all moved forward. If they failed, it would be lights out.

As you may have sensed, I was all in. I believed in Lloyd and everyone around him. I envisioned *Revelation II*. I imagined sitting in a movie theater watching a nuanced thriller, when suddenly, unexpect-edly, a larger-than-life-sized three-dimensional image of Christ would magically appear in front of the movie screen. Equally, I knew that the concept sounded spectacular, but if I didn't have a prototype to demon-strate its viability, no one would put up a penny.

On February 27, 1973, Lloyd Cross and I officially formed Optical Infinities, signing a partnership agreement.

> Tuesday, February 27, 1973—School of Holography
>
> Woke up and felt frenzied cause I never really slept. Got Pam and Lloyd up. Rain. Lloyd doesn't want to go downtown without

a car in the rain. I agree to wait. We work out a plan again. I'm so strung out I can hardly see. Lloyd has all the patience. I'm anxious to get home. We get draft of agreement notarized—I get on plane, fly home. Don Richetta calls saying he may have someone interested in investing (in the film).

Linda Lane
555 N. Bristol Ave.
Los Angeles, California 90049

27 February 1973

Lloyd Cross
454 Shotwell St.
San Francisco 94110

OPTICAL INFINITIES

PARTNERSHIP AGREEMENT

The purpose of this partnership between Lloyd Cross and Linda Lane is to set up a company to produce and sell white light transmission holograms using ambient light reconstruction. The initial products of the company will be based on the three (3) axis diffraction grating films currently being produced at The School Of Holography, located at 454 Shotwell St. San Francisco, California.

Lloyd Cross is the sole inventor and owns all rights to this product and he agrees to assign these rights to the company.

Linda Lane agrees to provide monies required by the partnership prior to the formation of the company and to pursue the acquisition of investment capital required for the company.

State of California
County of San Francisco } ss

— ACKNOWLEDGMENT—General —

On this 27th day of February A. D. 1973 before me, DAVID F. GALLAGHER Notary Public a Notary Public in and for the said County and State, residing therein, duly commissioned and sworn, personally appeared LLOYD CROSS AND LINDA LANE

known to me to be the persons whose names are subscribed to the within Instrument, and acknowledged to me that they executed the same.

In Witness Whereof, I have hereunto set my hand and affixed my official seal the day and year in this Certificate first above written.

David F. Gallagher

Notary Public in and for said County and State of California

DAVID F. GALLAGHER
Notary Public - California
City and County of
San Francisco
My Commission Expires Mar. 12, 1973

My Commission Expires

Form GA — Sam Hopkins Legal Forms Printing Service, 2328 Fruitvale Ave., Oakland, Calif.

Lucky for me, nine days later my parents were slated to attend the Associated General Contractors convention at the Fairmont Hotel in San Francisco. These were my father's peers—men who built skyscrapers, airplane hangars, homes, and highways—the big boys, from all over America. Their cocktail parties were as glamorous as the heady soirees depicted on *Madmen*.

Of course, nothing came without strings. If I wanted to hitch a ride with my parents, I had to agree to play the well-brought-up, daughter. I did what was required while my hippie heart pounded beneath my little black dress and cultured pearls. My actions were not up for negotiation. After each appointed event I changed clothes and took the bus across town.

Arriving at Shotwell St. was heaven. Everyone at the School of Holography was doing his or her part to build the multiplex camera. Lon Moore, the school's first student, having arrived from New York in the fall of 1971, was helping Lloyd make a rotating track

> Tuesday, February 27, 1973
>
> Waiting for the resin to set to see if 360 (degree) gear setup for multiplexor will work. After 1 am. Rain. A bright, diligent young artist-scientist discussing the possible steps for actuating the camera—Lloyd is the School of Holography. No one knows more about science, life and dreams than this magical man.

From my diary entry it's pretty easy to sense the master/student pool of awe.

The machine Lloyd was engineering would take the 16mm footage of Alice Cooper and, using a circular platform rotating at 24 frames per second, record it onto holographic film. Once developed and reconstructed using a laser directed from the same angle as it had been shot, a three-dimensional recreation of the image would appear to float in the center of the cylinder.

By combining conventional cinematography with holography, Lloyd Cross was inventing integral or multiplex holography. The Dalí hologram required 1,080 frames to be transferred onto 12-inch-wide holographic film. I couldn't wait to arrive at the beehive of activity to see what they'd accomplished.

> Friday, March 9, 1973—School of Holography
>
> I take the bus to the school. I meet Selwyn (Lissack). Tall, slight, large brown eyes with depth. Originally from South Africa. Musician's hands. He arrived last night. Energy level high. I find out that Gary (Adams), Sharon and Stonewall are doing a multi-media show in Chico. Got business out of the way. Took bus back through Haight. Paranoia personified.

Saturday I did my duty and accompanied my parents to various social functions. All I could think about was getting back to the school. What was I missing?!

> Sunday, March 11, 1973—School of Holography
>
> I went with Selwyn to pick up a camera and ended up talking for hours. I read Selwyn my poetry and "Controlled Environment" (a TV concept). He flips. He's going to call Cecile (Ruchin) in New York. He wants me to go to New York and work with them—and yes, I'm ready—if only there were 48 hours in a day.

The meeting with Selwyn Lissack opened Pandora's box. He and Cecile were 100-percent New York serious about the business of art and the business of holography and the business of making money. They were not hippies dazzled by 3-D.

Selwyn had grudgingly hired Lloyd to make the Salvador Dalí-Alice Cooper integral hologram, and if the project was going to succeed, he was going to have to be an eagle watching over his nest. From past experience and other people's stories, he knew he had to hover. Deadlines were always tricky, but this one was critical, and the last thing he wanted to do was disappoint Salvador Dalí.

With the concentration and zeal of a man herding cats, Selwyn was on a desperate mission to smooth the waters and create an environment that would help the group accomplish what many thought was impossible. The tension was palpable as Lloyd and his band of outliers worked long into the night.

College students and high school dropouts, artists and techno-freaks, the physicist-visionary and the hotshot promoter from New York worked day and night, night and day. Rolling cigarettes because they couldn't afford ready-made, drinking coffee after coffee to stay awake on crisp spring mornings. Jerry-rigging a system because they had to. You could feel the energy—the history making, once-in-a-lifetime excitement that radiated throughout the building. I took photos of the multiplex camera being built and as far as I know, I'm the only person who did.

The large half of an unpeeled carrot had been left on top of the master timing cam. The laser image projector, which looked like a camera, was the cylinder drive mechanism.

Timing had to be perfect. This was a make or break moment for everyone. The mood was slow. Voices were soft spoken. Everyone moved cautiously, carefully. No one took a break.

My first impression of Selwyn, originally from Cape Town, South Africa, was that he was charming, nice and patient. As time went on my impression changed. During this critical stage, Selwyn and I stopped speaking. To cut my own tension and do something useful, I took Polaroids of the building process.

Sawdust permeated the air and saw noise canceled out the voices. Hand-rolled cigarettes, a giant sack of carrots—hardworking individuals pulsing together to make the makeshift magic machine work.

Machinery in the midst of dirty glasses, equipment, string, hand saws, paper towels. The appearance was far from sterile or *top secret*, as in sterile government laboratories. The mood was quiet, methodical. No one's eyes seemed to notice the dirt or the confusion—only the machine that they are creating.

So here we were…waiting…praying that Lloyd Cross' genius would shine through and the first 360 degree multiplex hologram of Alice

HOLOGRAPHY

Holography: Greek meaning "whole message".

Laser stands for: Light Amplification by Stimulated Emission of Radiation.

Hologram: A hologram is a lensless laser photograph which has volume, depth and parallax in three dimensions.

When taking a photograph one selects the subject -- aims the camera -- adjusts for distance, light and focus, then snaps the picture. When making a hologram the laboratory or artist's studio is comparable to the inside of a camera. Each measurement: light, distance, shadow and vibration must be perfectly calculated. Otherwise, there will be a series of dark lines which reflect nothing. A hologram must be made with coherent light and the laser is the best coherent light source. The light waves emitted by a laser beam are identical in length which means that they have a regular frequency. To illustrate the set-up for making a hologram we have a diagram below. For more information please consult your local bookstore.

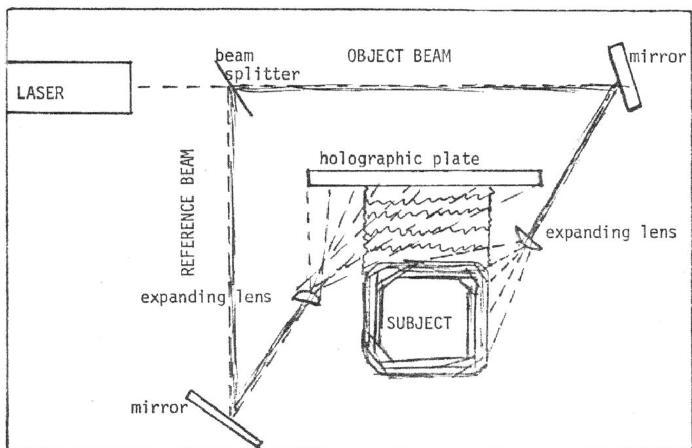

Cooper, brain with ants by Dalí, and wearing one million dollars worth of diamonds would appear to float in the middle of the cylinder.

It was the challenge that kept the grindstone spinning. We were ready to see the first test shots. I asked why no one was videotaping this historic event, and I was told that the video group next door was struggling and didn't want to spend the money on tape.

To incentivize the exhausted, Selwyn tells everyone that there's fifteen minutes of color videotape in New York that's going to be added to the hologram—Dalí performing magic tricks. Now who wouldn't like to see that on film?

There was no kitchen and everyone was hungry. Someone named Allan started making a giant salad, boiling eggs on a hot plate and cutting up apples. That wasn't enough for twelve hungry scientists, so cigarettes and hot black instant coffee filled the void.

It's Sunday night and there's no place to buy film. We have to wait.

Then there's the matter of the large plastic cylinder. It was discussed and agreed that it would take a local company at least ten days to fabricate. It would cost a hundred dollars, and the manufacturer wouldn't hurry because the job was for Salvador Dalí.

> Monday, March 12, 1973
>
> Selwyn and Cecile want me to move to New York to stay with Cecile and write the Holographic Journal. I'm ready—no home, boxes packed—ready to move. Selwyn is very together. Offer sounds good and I hope I can do it.

Walking past the multiplex machine I heard it hum/stop/hum... go on instrumentally as it waited for the rest of its parts to be attached. One person was working the laser, another the mirror. All parts were being carefully fitted. The plastic cylinder was being molded. What would take a major company a month, a team of experts and a hundred thousand dollars was costing just a thousand and the devotion of a few skilled and unskilled techno-freaks.

Someone said, "You know why Lloyd can do anything?" I shook my head. "It's because when he was a little boy he'd ask his father to help him with something and instead of telling Lloyd to go away, he always attempted to find a solution, a way to make it work."

That was exactly what I was witnessing. There were six young men trying to build a machine using their imagination and ingenuity; their hands, power tools, large saws, and a hot plate. Necessity was still the mother of invention.

Monday, March 12, 1973

Selwyn's mentioned that he'd like to play Jay (artist in Revelation). He really, really likes the script. One rewrite. I have a cold and have to get back to L.A.

Tuesday, March 13, 1973—School of Holography

Went back to School, cold and all for the test shots. Went with Selwyn to buy food. He told me never to eat tomatoes because they're so acidic they'll make your hair fall out. Got food at Zim's and went back. Selwyn offered $35 per person to work all night. Yes, all would.

It had been forty-five days since the building process began. The track was set, the glue, dry. It was time to find out if Lloyd's theories and calculations would produce a moving holographic image. The motion picture film that Dalí and Selwyn shot in New York was ready to be loaded onto the circular, rotating, jerry-rigged mechanical device.

Wednesday, March 14, 1973—School of Holography

Parents late so I plead my case. Stopped at SOH and saw the first multiplexer test plates ever made. There was Alice Cooper suspended mid-cylinder next to Christ. Floating krypton speckled depression glass green (laser light). Historic art! The machine worked. Lloyd, the genius and Selwyn, the artist-supervisor.

On the drive back to Los Angeles I was elated. Everything was coming together. Dalí would open a new door into his visual brilliance and the art world would explode with enthusiasm. The publicity alone would make financiers and studio brass less skeptical, and more daring. Of course, seeing a prototype that had to be illuminated by laser light was only a baby step, but all the same, it was a step in the right direction.

CHAPTER 7

STRIKING A BALANCE

When I arrived back in L.A. a celebrity I knew called to ask if I'd be interested in ghosting a few chapters of an autobiography. There were two considerations: first, a large sum of money, and second, a confidentiality agreement. At this point, money was speaking my language.

My mood was manic-depressive. One minute it looked like my film was moving forward, and in the next external forces were blasting everything into another set of obstacles. Challenges are not unusual when making a film. That said, when you're trying to invent a new dimension, if it works, you're George Lucas, and if it doesn't, you're dust.

> Saturday, March 17, 1973
>
> So fucking depressed can't think straight. Know that Gilbert and Richetta are overreacting because of personal lives, but I feel drained when talking to Charlie (Patton) who tells me about David Bowie people wanting a life-sized hologram and him telling them how they can do it. That "yes," it can be done. I'm praying that the call at 6 will bring good news from Lorimer (Gilbert and Richetta's attorney).

Once again I found myself waiting for someone to complete

another piece of the puzzle. When an audience watches a film, they're transported to a very specific world. No one considers the number of individuals it took to create and maintain the continuity of that special cosmos. As a screenwriter I was extremely frustrated.

The Writers Guild was still on strike, so I wasn't supposed to rewrite anything. I was doing my due diligence and picketing while listening to Bob Gilbert tell me I needed to ignore the strike and rewrite the screenplay for the good of the project. I felt tremendous pressure, and it wasn't a good feeling. I was deeply conflicted. On one hand I envisioned a new, three-dimensional, interactive way of experiencing motion pictures. I wanted to make *Revelation* better. On the other hand I felt bound by the rules of the Writers Guild.

My father, a building contractor, had told me horror stories about pre-union carpenters who went from town to town using their skills to build a building only to have the boss run them out of town without payment. My father was a Reagan Republican with great respect for unions. Lest we not forget, Ronald Reagan had been president of the Screen Actors Guild. I knew in my heart that if there was no Writers Guild, writers would always be at the mercy of studios and networks. There would be few paid rewrites, and no residuals.

Holography had to be the wave of the future. The more I thought about it, the more I envisioned cities illuminated by streetlights projecting spectral rainbows, and holographic destinations embedded in sliding glass doors. Press one for a sunny Caribbean beach, two for an Amazon rainforest, ten for the Swiss Alps. Escape would be at one's fingertips. The future of holographic innovations seemed infinite, and weighing the two paths, screenwriting verses holography, I was magnetized by the latter. I was beginning to feel like Laser Lady.

Sunday, March 18, 1973

Jenny (Corey) and her friend Dick took me to Anait's showing. A disaster! Combination home movie—entertainment combining colors with sound. Interesting concept til Anait came out in white Afro wig, leotard, tights, and make-up with matching male partner. Terrible experience.

Certainly not everything Anait Stephens did worked. She was brilliant, rich, and persistent, and her pioneering efforts would ultimately override her less impressive moments of rapture. She was determined to be recognized as a consummate artist.

I was determined to complete my picketing assignments while ghosting chapters of someone else's autobiography. Books didn't fall under the auspices of the Writers Guild, so I was safe. The job paid well and that was all that mattered.

At the crack of dawn Monday morning I held my sign high at Paramount from 6am to 9am in the hope that someone would come to work early and realize that we were committed. We were a stubborn, outspoken group willing to sacrifice whatever it took to make professional screenwriting a career with good benefits and financial parity. Walking the line gave me plenty of time to think, evaluate, ruminate... Think, talk, walk...think, talk, walk...

By the time I got home I was exhausted. I closed my eyes and considered the possibility of working with Selwyn Lissack and his partner Cecile Ruchin in New York at HCCA. I would be their West Coast representative. There was nothing stopping me from writing the holography journal they were proposing, but I needed to go to New York to solidify the arrangement.

I wrote Cecile, sharing my enthusiasm, and capping the letter off with ten photos I'd taken at the School of Holography featuring Lloyd Cross and company building the first Multiplex machine.

Selwyn and Cecile were successful entrepreneurs. They had convinced Quaker Oats to offer the world's first holographic premium, the King Vitamin hologram ring.

This was huge. A major corporation was backing a red plastic ring with King Vitamin appearing to float off its round surface. These were reflection holograms produced by McDonnell Douglas, a *Fortune* 500 corporation known for manufacturing jet fighters.

I discovered that in 1966 McDonnell Douglas–Missouri had purchased an ailing Ann Arbor, Michigan holography company called Conductron. McDonnell Douglas was business as usual—dry and concise. Conductron was made up of physicists and engineers with a think-outside-the-box attitude and a proclivity for art. New blood would give McDonnell Douglas an edge in a burgeoning new field.

Conductron became the beneficiary of deep pockets and state-of-the-art lasers and equipment. They could produce prohibitively expensive pulsed laser portraits of living, breathing subjects. General Motors and pharmaceutical giant Hoffman-La Roche commissioned holograms for trade shows, and in 1972 artist Robert Schinella worked with Conductron to produce "Hand In Jewels," a real human hand dangling a diamond bracelet, which appeared in front of Cartier's Fifth Avenue store window. It was a spectacular illusion that brought crowds, along with the world press.

In a handwritten letter from Cecile Ruchin dated January 29, 1973, she writes:

> Dear Linda,
>
> So much is in motion now that time to fill you in is not adequate in writing.
>
> All plans on corp. HCCA going ahead—have $100,000 financial proposal for Jewelry being completed with a man who's out raising the $$. When finished you will get a copy. If you rep you get 5 percent finder's on all deals.
>
> Have Dolgoff teaching course (March) called "Reality—from Perception to Holography." Have applied for grant for med research in holography diagnostics. All proposals are looking like they will go through. Am in touch with Holoconcepts and we can use Schinella M'Dec + not Lloyd.
>
> Lloyd still floundering and creating lack of confidence by his refusal to do jobs on schedule and when promised.

> Our man Lou Brill will contact you when he arrives L.A. (sometime within next 3-4 weeks) Also—Dalí is back and we have action and real potential sales on Multiplex but afraid to move with Lloyd under circumstances. (There are alternatives it seems—will tell you more as it develops).
>
> Love, Cecile

It was now the end of March, the strike was still on, and I was determined to get to NYC where it was all happening. I'd seen enough at the School of Holography to know that Cecile's trepidations were well founded. I'd also seen enough to know that this was the opportunity of a lifetime.

That weekend Henry Seldis, the art critic for the *Los Angeles Times*, had invited his close friend, my artist-mentor Channing Peake, to a wedding in Malibu. Channing invited me, and although our romantic relationship had cooled, I loved spending time with him.

Channing was the ultimate gentleman cowboy. Living on a ranch in the Santa Inez Valley had given him the peace of cattle country and the sophistication of Santa Barbara and Montecito. His wardrobe consisted of jeans, cowboy boots, well pressed shirts, beautifully crafted silver and turquoise Indian jewelry, and Stetson hats. Irish DNA fed his outrageous sense of humor, especially after a few drinks. Even at sixty, Channing maintained a mischievous twinkle in his eyes.

Frank Perls was Channing's art dealer and close friend. A tall walrus of a man with a tobacco-stained mustache and yellowing teeth, Perls was as sharp and roguish as the artists he represented. Picasso was his most famous client.

The two bon vivants loved to tell the story of the time they were visiting Picasso in the south of France. They were at his Mougins studio when Picasso fashioned a small bull out of clay. Of course, Perls and Peake wanted a bronze of the little model.

Picasso, the shrewd businessman, made them a gift and a deal. They could have ten pieces struck—a limited, numbered edition. Picasso would receive six tiny bulls while Perls and Peake would pay for everything and each receive two. The mold would be returned to Picasso.

Being with Channing and having him share his world of contemporary art with me helped mold and inform my passion for collecting. I believed the connection between Salvador Dalí and holography as an art form was indisputable, and I felt that I was in a prime position to convince Henry Seldis that the L.A. County Art Museum needed to use "art" and "hologram" in the same sentence.

Certainly, these three-dimensional works of wonder had their limitations. Lighting and positioning were critical, and if you didn't get everything just right you weren't going to be able to see an image.

In the early 1970s, with few exceptions such as Bruce Nauman, holograms were novelties. No one knew if holographic emulsions would stand the test of time.

There were more questions than answers, but it didn't deter me. Henry Seldis was a powerful voice in the L.A. art community and I had his ear.

Fate...divine providence...call it what you like—I felt like I was in all the right places at all the right times. History was being made, and I was part of it.

CHAPTER 8

NEW YORK, NEW YORK

Monday, April 9, 1973

3 to 6 Paramount. Today it's overcast. Feeling strange about Selwyn and Cecile. Maybe he was putting me on. I wait. Go to the office. No calls. Late for picket line. No one says anything. Eight or so the phone rings. It's Selwyn and Cecile. They want me to come to New York. All is peaceful.

Wednesday, April 11, 1973

Preparing for the mythical mugging New York streets. Doing everything for battle. Going armed with literary material and charm, new clothes and the approach of Spring. Having no idea what to expect. Not knowing where I'm staying, what they'll be like. Just following one gigantic, urgent gamble. An intuitive need to spend two weeks in New York City.

In 1973 New York was gritty and electric, living up to its name: the city that never sleeps. To me Manhattan epitomized the highest rung of the talent ladder. Be it publishing, theater, art, advertising, or fashion, if you were respected in New York you had to have earned it.

As for me, I was every inch a California girl with stars in my eyes. I was convinced that as soon as the Writers Guild strike was over, I would be financially independent. I was confident about everything until my friend Charlie Patton started painting an unsettling picture of Manhattan. I'd only been there twice, once after graduating from USC and once with my parents. Both times I stayed in uptown hotels with a tightly scripted itinerary. Now I was being warned that nice, safe neighborhoods could suddenly become dilapidated, menacing danger zones. "Stay alert! Pay attention," was Charlie's warning.

Taxis were expensive, subways made me insecure and claustrophobic, and even though I felt optimistic about future earnings, I opted to economize and stay above ground. That meant walking, taking buses and learning Manhattan's street grid.

> Saturday, April 14, 1973—New York City
>
> Flew in a 747 with a lounge in back. Five-hour flight. Good, met interesting, nice people. Still did not dispel my innate New York paranoia. Arrive at Kennedy at 5:30 pm—take a bus into NY to save money. Take taxi from upper Park Avenue to 865 Broadway. Strange area. Rather deserted. Storefronts fenced for weekend. Cab driver appears concerned...waits as I go to gate and pull— it's open, not locked. I walk into a dark hallway and press the elevator button. The elevator can be heard clanging its way down. As the wooden door swings open I see a pixelated lady who smiles and extends a firm hand. She is Cecile Ruchin, my hostess, a woman engaged in entrepreneurial battle and showing signs of that three-year struggle. Her face is angular. Strong bone structure. Rail thin. Deep circles, warm smile. We go up in the elevator. Selwyn is watching TV.

Cecile's loft was warm and immediately inviting. I'd arrived armed with my friend Joan Nielsen's album cover, artist Robert Blue's vision for *Laser Lady*, and my poetry and writing samples. It was a night of reverie, celebrating a future that would both change and advance the way everything from X-rays to movies were seen. In two weeks Lloyd Cross would be unveiling integral, also known as multiplex holograms, at the International Symposium on Holography. Joe Greenberg, Alice Cooper's manager, was setting up a huge press conference to introduce New York and the world to Lloyd's new three-dimensional technology. We were beside ourselves with anticipation.

The loft was divided into a number of living spaces as Cecile had two children: a beautiful sixteen-year-old daughter, and a younger streetwise son. There was never a dull moment.

Selwyn had his own apartment, giving him the space he needed. Clearly the South African drummer was a thinker. Tall and silent much of the time, he listened and assessed all the moving parts of past, present, and future efforts.

My first night at Cecile's was heady. We were all on the same page. HCCA had been in business for three years. They had scaled a mountain of skepticism to raise enough capital to establish themselves as pioneers in holography. I was there to write the holographic journal, press releases, proposals, whatever they needed.

On Sunday, Cecile took me to brunch at her friend Henry Madden's, an actor whose mellifluous voice had made him the gold standard for voice-overs. He and his friends, who were all in the arts, made me feel at home. In New York, I had a sense of belonging—of being connected to kindred spirits in a way that I had never experienced in L.A.

In order to grasp the lay of the land, Cecile took me to Central Park and Greenwich Village. I felt like we had known each other for years. She had a unique futuristic style, always wearing a jumpsuit with a variety of cloche hats covering thin wisps of light brown hair. I decided I was being guided by an omniscient creature.

> Monday, April 16, 1973
>
> Went to Alive Enterprises, Alice Cooper's manager's office. Met Joe Greenberg (Alice's manager), left (writing) material with him, called Peter (Cookson) and made arrangements to meet

> for lunch. Went to mediocre Italian restaurant near Alive, then on to Museum of Modern Art. Went back to Alive for meeting that didn't happen, so went home. Afraid to go out alone at night cause near Washington Square.

Ironically, Cecile's phone had been turned off and accidentally cross-connected to a box company. This meant that we couldn't use it during the day, but once the box company closed—nights and weekends—we were in business.

Tuesday I was supposed to meet with Joe Greenberg about writing the holography journal, but he put it off. I got the feeling that he was completely committed to all things Alice Cooper, and if that meant a small foray into holography, he'd be a good sport and humor us. It was disappointing.

I made a date with my former mentor, Peter Cookson, to see *The Championship Season*—the Broadway drama that would win the Tony for Best Play in 1973.

Afterwards Peter took me to Max's Kansas City, the gathering spot for Andy Warhol, Betsey Johnson, Robert Rauschenberg, David Bowie, Patti Smith, Sam Shepard, and many other artists, writers, filmmakers, poets, and musicians. It was both a restaurant and a nightclub. Young artists like Bruce Springsteen, Bob Marley, and Willie Nelson played there. It was the Troubadour on steroids.

The evening ended early, and when Peter Cookson dropped me off at Cecile's loft, I realized that I was staying a block from Max's. I called the club and asked for Warren Finnerty, an actor and friend of Charlie Patton's whom I'd met in L.A. Miracle of miracles, he was there. I told him where I was staying and he offered to walk over and escort me back to the nightspot.

> Monday, April 16, 1973—Max's Kansas City
>
> Big as life, Richard Harris stands out amidst a cluster of artistic looking people. Richard is telling a curly haired girl a story.

Warren and I pass on our way to the tables in the rear. A conversation crosses my mind: I'm a poet, Richard. A good poet! To which he responds favorably and we establish a relationship.

Warren introduces me to Milton Ginsberg, a filmmaker who's just finished A Werewolf in Washington starring Dean Stockwell (small world). I convince him that I'm a serious writer looking for a NY agent and he agrees to help me, but he wants to split.

I go upstairs to see what's happening—dancing. No Warren. I think of Richard Harris and go downstairs. He's still there. I find Warren. A nice Jewish man, 40ish and hip begins talking to me and buying me drinks. Before I know it I'm tipsy and Richard Harris is standing next to me.

I'm looking at the tall Christ-like figure when the man I'm talking to says, "IRA." Richard reflexively snaps, "What'd you say about the IRA?!?" Man says nothing. I jump in, "Oh, you're in New York to publish your poetry, aren't you?" "Yes, that's right!" he says. "I'm a poet, a good poet and I'm in New York to see about getting mine published."

During this I'm nervous and he's continually distracted. Drunk.

He says, "What're you doing tomorrow?"

"I don't know."

"Let's read poetry tomorrow! When?"

I realize he's drunk and has to sleep late...but he'll probably have a meeting in the afternoon...

"How's twelve o'clock?"

"Twelve. Fine!"

"Where?"

"Algonquin."

I think of the legendary literary Algonquin Round Table with Dorothy Parker and Alexander Woollcott. Knowing where to find him is essential. "What room?"

"905."

"You won't forget? Twelve tomorrow?"

To which he turns and repeats the time. I went back to my other conversation secretly elated.

A fight had been averted, and in the process I'd been rewarded with an invitation to read poetry with Richard Harris, one of the most talented men on the planet.

Wednesday, April 18, 1973

I couldn't sleep. Was much too worried and excited. Got up early and decided what to wear—moss green suede shirtwaist dress, no purse—leather briefcase filled with material and a few holograms. Tell Cecile in case I don't come home so she won't worry.

I take the bus to 46th St. Walk two blocks—see Algonquin. It's 12. I enter the antique edifice—don't see the elevator and don't want to look lost so go to the ladies room. I ask a woman, "Where's the elevator?" She doesn't know. I spy a stairway. First floor and there's the elevator. Nine—I get off—walk to 905 and knock.

The door is opened by a bearded, rotund elfish man of forty that I remember from the previous night. "Is Richard Harris here?"

"Yes," he says taking me in from head to toe. "Do you have an appointment with him?"

"Yes, at twelve o'clock."

He disappears and reappears in a flash leading me inside the suite and past two other men in the sitting room. I'm ushered into the bedroom where Richard is vaguely covered with only a white sheet. He looks every bit the movie star with mischievous, inviting blue, blue eyes.

"Hello pretty," he says, taking my hand and moving it towards him. "Take your clothes off and get into bed."

"I don't give public performances!" I shot back, pulling my hand away.

"Oh, they'll be gone in a minute." I stood my ground, internal pyrotechnics taking over. I could hear the three men who I later found out were Richard's brother Dermot, Terry James, who I thought of as Terry the elf, and Richard's manager, Syd. Just as Richard promised, the voices trailed off and the door to the sitting room slammed behind them.

I'd seen the Irish actor play King Arthur in *Camelot* and John Morgan in *A Man Called Horse*. I'd heard him sing "How to Handle A Woman" and "MacArthur Park."

The fourteen-year difference in our ages mattered not.

I'd grown up around larger-than-life men. My own father had been California State Skeet Champion and a world record fisherman. Wendell Corey, my best friend's father, was a movie star who'd been president of the Academy. I was used to being in the presence of movie stars, but this was different. I was a delirious fly caught in a honeyed web.

Eventually Richard's entourage returned and toast and tea was ordered. I took out a couple holograms, gave a brief introduction, and passed them around. Without a proper light source such as a mercury arc, a laser, or the sun, seeing the full three-dimensional effect was hit and miss. The holograms were received with interest and dismissed as a novelty. Richard asked me to read my poetry aloud. Fortunately, that went over better than the holograms. I began:

Big houses and empty heads
Bouncing off the walls,
Big Houses and empty ears
Waiting for telephone calls.
Minds of knowledge
Lost in space
Hearts of passion
Locked in place.
Souls once fertile
Dry with age
Innocence once open
Now kept caged.
Big houses and empty heads
Waiting for a cause,
Big houses and empty ears
Begging for applause.
Thoughts of passion are
Canceled by dawn
As crèmes of rejuvenation
Try to prolong
Big houses and empty heads
Bouncing off the walls,
Big houses and empty lives
Pacing through the halls.

There was applause. Richard called me his "Lovely Lascivious Laser Lady" after another set of my poems and told me how much he liked my poetry. I didn't want to leave, however I had an appointment with an agent at Curtis Brown, and Richard had a two o'clock meeting.

"Lascivious Lady, meet me at the Blue Bar at three," and he was gone.

I apparently found the literary agency, but I have no recollection of what was discussed. I speed walked the mile back to the Algonquin and found Terry, Dermot, and their friend Richard Okon having drinks in the Blue Bar. They were a witty, naughty, gleeful group and I had to keep up.

Richard was narrating the enormously popular book, *Jonathan Livingston Seagull.* Terry James was composing the music. Dermot kept a close eye on Richard's business affairs.

"Do you have money? Are you rich?" Dermot, the blue-eyed Scorpio started prodding me.

I chose to play his game and say nothing one way or the other. "I hope I'll always have more sense than dollars."

Shrieks of approval. So far so good. I ordered a drink and started reading palms. This was a skill I'd picked up from an old book I'd discovered in the Coreys' library.

Finally Richard and two men from the Sol Hurok organization arrived. He greeted me with a hearty "Hello pretty!" and ordered a rum and ginger ale. The conversation turned to Picasso's recent death and estate.

At 5:30, Richard, "T" as Terry was called, and I went to the Algonquin Bar for dinner. Richard was demonstrative, holding my hand under the table, complimenting me. Oysters on the half shell, filet mignon, creamed spinach, petit pois, and drinks. Lots of drinks. Richard told stories about his three wonderful sons, his early starving actor days—and Joan Littlewood, the woman who discovered him and gave him his first starring role.

We visited a Ukrainian gift shop where Richard dug through a pile of LPs, found his recording of *Camelot*, exchanged it with the album on the turntable and dazzled everyone within earshot. "What would you like?" he asked, motioning towards an eclectic assortment of handmade reminders of someone's old country. I chose a beaded egg, an elaborate gilt frame, and a small art piece. It was both exhilarating and surreal. Was I on the inside looking out or the outside looking in? I wasn't sure about anything beyond our symbiotic chemistry.

The taxi ride to McSorley's, the oldest Irish bar in the city, was *Fifty Shades* exciting; dangerous and romantic. Richard insisted on taking me there. One round of drinks and it was off to the Mad Hatter, a raucous

Irish bar where we were joined by his entourage. It was an over-the-top evening of great fun mixed with anxiety.

There was nothing subtle about Richard Harris when he became drinking buddies with the boys at the bar. I watched several beer-soaked instigators itching to take on the man with a reputation for spontaneous combustion.

I wanted to go back to the Algonquin. Richard wanted to continue drinking.

What seemed like thirty drinks later we made it back to the Blue Bar and finally upstairs.

I'd grown up around alcoholics so I knew the signs. In fact, that could've been one of the reasons I was so comfortable with him. We were supposed to see Rex Harrison in Pirandello's *Henry IV* the next night, but the binge had finally taken its toll. Rex Harrison, then married to Richard's ex-wife Elizabeth, added to the actor's angst. The two men weren't exactly friends. In fact, I don't know why Richard even considered going in the first place. He was off to Durango, and I was extending my time in Manhattan. We vowed to see each on the West Coast when he filmed *99 and 44/100 Percent Dead.*

> Monday, April 30, 1973
>
> Joe Greenberg hires me to write a 10-page piece (on holography) for $300. He's skeptical. I have until Thursday. We're still waiting for Lloyd (Cross) to arrive. I begin writing. I'm going nuts over holography and integral holography. Don't know if I can do it.

My job was to translate a scientific process into layman's terms while making it sound interesting and exciting. I was used to writing scripts with dialogue, not technical papers. Joe Greenberg and Shep Gordon had formed Alive Enterprises. They were heavyweight managers in the music business. Joe was wiry, aloof, and confident. He came across as an impatient New Yorker, definitely the man on the decision-making side of the desk.

Wednesday, May 1, 1973

Wednesday night Lloyd (Cross), Pam (Brazier), Michael Krassner, and Lon Moore arrive with integral holograms. We're all on top of the world. We're counting our press releases before they've been written. I attempt to glean info on Integral Holography (Multiplex) from Lloyd. We all want a precious moment (with him). Pam looks great—all curls, angora sweater. Michael Krasner is Lloyd's twenty-six-or-seven-year-old attorney. Straight, does yoga exercises morning and night.

Lloyd Cross and company had flown in from San Francisco to present a paper on integral holography at the Bio Medical Conference. Lloyd's mission was to convince scientists and doctors that his holographic advancement—shooting motion picture film that could be transformed into a three-dimensional, 360-degree, real-time moving image—would improve everything from x-rays to medical research.

Lloyd needed to show this illustrious international group of corporate and government thinkers that his latest invention worked, and that it urgently needed to be funded. As I mentioned earlier, Cecile had written grant proposals for this very purpose. Lloyd Cross' presentation would be the critical, deciding factor.

The guests of honor were Dr. Dennis Gabor and Dr. Emmett Leith, the fathers of modern holography, and Dr. Pal Greguss, the inventor of Acoustical Holography and the symposium organizer.

In 1971, Dennis Gabor, the British-Hungarian electrical engineer and physicist, won the Nobel Prize for Physics "for his invention and development of the holographic method." In 1947, while watching a tennis match, Dr. Gabor envisioned the concept for making holograms, and one year later, using a mercury arc lamp with a narrow band green filter, made a crude example. He coined the term "hologram"—*holos* meaning "whole" and *gram* meaning "message." Unfortunately though, it wasn't until the laser was invented in 1960 that a coherent light source became available.

In 1964, Dr. Gabor was teaching and inventing at the Imperial College of Science and Technology in London when, across the pond, Dr. Emmett Leith, a professor of engineering and computer science at the University of Michigan, Willow Run Laboratories, and his colleague, Juris Upatnieks, a physicist and inventor, showed the world's first clean three-dimensional hologram at an Optical Society conference.

Suddenly, the crude, double-image experiments Dr. Gabor made were a thing of the past. The coherent light of the laser allowed Leith and Upatnieks to make a clear three-dimensional hologram of a train. Interest in lensless laser photography hit the world stage, and since Dr. Gabor had documented his early findings and experiments, he was crowned "The Father of Holography."

Dr. Gabor's comment upon seeing the first integral or Multiplex holography display at the symposium was "Splendid." Of course, this pleased its architect.

Lloyd told me that in the summer of 1972 he made the first integral holography test using thirty photographs of a young woman. The results indicated to him that complete success could be achieved by building a machine that would transfer the images onto holographic film using a laser. A special machine would have to be built to his precise specifications.

The first integral holograms were called *cronograms* by the people shooting them at the School of Holography in San Francisco. Lloyd arrived with a self-portrait, a group shot of him with his co-workers, individual portraits, dancers, and the first 360-degree 3-D x-ray.

Referring to my original papers, I found the following paragraph:

"These integral holograms are now on public exhibition at 1 Park Avenue, ground floor, New York City. This realizes his (meaning Lloyd) original commitment to the New York State Council of the Arts by the creation of a new visual medium."

Obviously, Lloyd had been funded to create a new visual medium, and almost by accident he had finally delivered.

By Thursday, the need to conquer the technical writing beast coupled with the importance of being paid kept me at the typewriter around the clock. I was seriously disappointed to miss Dennis Gabor and Emmett Leith's presentations at the medical conference.

Then Joe Greenberg gave me an extension on my papers along with two tickets to *Pippin*, the hottest ticket on Broadway. I was supposed to consider Ben Vereen, the musical's star, for Father Fantastic, the charismatic black preacher in *Revelation*.

I took Richard Okon, whom I'd met through Richard Harris, to *Pippin*. Third row center. Nice. We sat through the show and during the end monologue Ben Vereen played, "You know, you're an exceptional person..." directly to me. Tired as I was, I had to meet him. Richard and I went backstage and Ben Vereen, the biggest star on Broadway, turned out to be one of the warmest, nicest people on the planet. The following month he won the Tony Award for Best Actor in a Musical.

I showed him a couple holograms and he became animated. He got it. He saw the future. Wow! After the lukewarm reception from Richard Harris' circle, I was reinvigorated. Ben Vereen gave me his phone numbers and told me to keep in touch. *Revelation II* was another step closer to becoming real.

Friday, May 4, 1973

I said the hell with it and went to New York Medical College to hear Lloyd (Cross) present his invention, integral holography, to the medical and holographic community. Lloyd in shoes with holes, and fragrant t-shirt, was unprepared. I bought him transparencies and grease pencils at the bookstore. He jotted notes and a funny, abstract poem for his elite Brooks Brothers peers.

The mature leprechaun was successful, but not as heralded as I expected. Had lunch—horrible hospital food. Rain—pouring. Must get equipment out by three cause they're locking that door. Guard waits as Selwyn tries to find Dr. Greguss. He brings the car around—we load large equipment and make a mad dash downtown picking up Lloyd and Pam on the other side of the hospital.

Reach Park (Avenue) and Lloyd wants the x-ray out of the suit-

> case. Suitcase? The suitcase containing all of the integral holograms and Lloyd's notes. We thrash around the station wagon, searching, lifting, not finding it.
>
> Gone. We've left it on the steps, on the street outside the locked door. I suggest a major publicity stunt—call TV stations. After all, this is New York. Leave a suitcase on the sidewalk and whap—it's gone in ten seconds.

The first Integral holograms looked like blank film until the image was reconstructed using a laser. This meant that anyone finding the suitcase and thinking they'd struck gold was going to be supremely disappointed. Once opened they would be unrolling what appeared to be pieces of clear plastic. Lloyd's notes looked like gobbledygook.

There were no cell phones. If you needed someone from the hospital to take a long walk outside you had to use a pay phone. Someone said they'd look, but we weren't entirely hopeful. At 1 Park Avenue, HCCA's home base the room was boiling with ideas.

I saw this catastrophe as a stepping stone to "News at 6." Has anyone seen an old leather suitcase filled with what look like clear strips of plastic and a scientific notebook? That was my version of making lemonade from a giant lemon.

Lloyd calmly began calculating how long it would take for San Francisco to make another set of integral holograms. We waited while someone at the hospital did our bidding. Traffic was bumper to bumper, putting us a good forty-five minutes away.

The phone rang and it was a breathless individual telling us that, "Yes!" he had our suitcase. Unbelievable as it was, all the brainstorming and calculating had been for naught. The suitcase was still sitting on the sidewalk in front of the hospital. It was a miracle and miracles were meant to be celebrated.

High on the positive outcome of our misadventure, Selwyn, Cecile, Lloyd, Pam, Lon Moore, and I joined Peter and Ana Maria Nicholson at Finelli's in SoHo for dinner. Both Peter and Ana Maria Nicholson were fine artists who pioneered 3-D lensless laser photography into a museum-quality art form.

Friday night we celebrated like it was New Year's Eve, all of us believing that fate had opened some huge golden door and invited us all to walk through it.

Naïve? Maybe. But exhilarating, definitely. I was so energized that first thing Saturday morning I headed to the New York Public Library—one of my favorite haunts. They had five books on holography and thousands on photography.

Sunday, Monday, and Tuesday I worked on the holography journal pieces for Joe Greenberg. There were no written contracts, so getting paid was up to him.

I continued editing and polishing twenty-five pages: "The History of Holography, Integral Holography, 3-D Photography," and a biography of Lloyd Cross, all to be submitted the next day.

> Wednesday, May 9, 1973
>
> Spoke to Fred Duncan (publisher) who said that he was very impressed with me and wants me to be the editor of the Journal of Holography. He gave me enough strokes so I said yes without agreeing to money and a real contract. I explained that I'd write him so we could agree (on terms).
>
> Went to Holoconcepts—saw Cartier hand holding diamond necklace from McDonnell Douglas dangling in front of the plate. I was very impressed. The hologram was made using a pulsed laser. Very, very expensive. Mind bending visual.

Wednesday night a group meeting was called to discuss financing Lloyd Cross's patentable invention. The group: Lloyd, Pam Brazier, Michael Krasner (Lloyd's attorney), holographer Lon Moore, Selwyn Lissack, Cecile Ruchin, and I met with the savvy holder of purse strings, Joe Greenberg. It did not go well.

Lloyd's transition from physicist/inventor to holographic artist/ inventor had given him the personality of a newly minted rock star, only without the trappings.

After a short time of discussing the upcoming press conference, Lloyd got hostile, asking Joe what he expected to bring to this project besides money. The irony of this being that it had been Joe who'd set up the press conference and planned the roll out. Yes, he'd put up money, but that was only part of his contribution.

Greenberg started to squirm. Lloyd told him that his creation of Alice Cooper was the lowest of the low and he didn't know if he even wanted to be connected with it.

Was Lloyd crazy?! He was killing the deal. I couldn't stay silent on the sidelines.

"If he could promote Alice," I blurted, "he could promote anybody! He could make peanut butter out of margarine!"

The mood lightened a bit and Lloyd explained that he needed $5,000 RIGHT NOW to pay the school's rent. Joe promised to spend $100,000 if that's what it would take to promote integral holography. In that moment Greenberg's commitment was strong and positive, but by the end of the meeting, once again, everything had cooled. He didn't write a check; instead he said we'd meet again on Saturday.

Thursday I'd finished the twenty-five-page treatise and I wanted to turn it in. Lloyd told me to hold off, it might need polishing. I acquiesced, reminding him it was due Friday, and I needed the $300.

"Joe hired me and Joe has to pay me. It doesn't matter if he uses the material or you use the material."

Friday, May 11, 1973

Lloyd continues to say "hold off." If I need money so badly, he'll get it for me—knowing damn well he won't. I owe my loyalties to Lloyd so I agree to hold off to ensure that he takes all my original copies to 1 Park Avenue. Fortunately, I make a few copies because ultimately, the envelope with all that material was thrown out or

> lost forever. I lose $2.76 (Xeroxing), $24 for Pippin tickets and Lloyd and Pam and Lon and Michael move out of Cecile's.

Saturday we met at 1 Park Avenue with high hopes. The ebullience slowly fizzled as we realized that Joe Greenberg wasn't coming. I gave copies of my written material to Alice Cooper's PR man. Clearly, no checks were being handed out.

Lloyd and his attorney thought they had an ace up their sleeves with two major corporations in the wings. They didn't respect Joe Greenberg's important role in producing the Salvador Dalí/Alice Cooper hologram or his business acumen.

Greenberg didn't need Lloyd; Lloyd needed Greenberg, but he was too full of himself to know it.

In a letter to Alice Corey I wrote:

> Lloyd Cross, genius holographer, inventor of integral holography, was supposed to come back to N.Y. for a major press conference. He was several weeks late. It was Alice Cooper's manager who was putting up the money to launch integral holography. Lloyd didn't like the idea of a snake charmer sponsoring his pet project. And, ultimately, Lloyd blew the whole thing. What was to be a parade of newspapers, TV stations and magazines turned into a one-minute fizzle. It was good to know this however.

I was hired to write the history of holography, integral holography, 3-D photography and a biography of Lloyd Cross. I was first given three days then extended to one week in which I literally worked 24 hours a day some days.

And when I was finished, Lloyd blows it with our money source who in turn doesn't pay me. It was an incredible experience from every point of view.

Instead of spending two weeks in Manhattan I was spending four. Selwyn and Cecile were proving to be entrepreneurs with business acumen. Unfortunately however, their alliance with genius Lloyd was fraught with too many specious, counterproductive encounters. Now they'd have to search for another physicist/holographer who was easier to work with and could meet a deadline.

I spent the next few days recovering. I visited my new friends—Henry Madden, the actor with a ceiling high cross in his living room and stigmata on his palms, and Richard Okon, who took me to brunch at Maxwell's Plum, where we discussed our friend Richard Harris ad nauseam. I saw *The River Niger* on Broadway and loved it. I followed up on publishing leads and even did a little shopping. I had spent an amazing month in Manhattan, and by the time I boarded my flight back to L.A. I felt like I was becoming a real New Yorker.

In that same letter to Alice Corey I also wrote:

> All in all I had more fun, work, and romance in New York than I have here (L.A.) in six months. I stayed with friends in a loft on Broadway between Park and Fifth. It provided me with total independence and I grew up in a hurry.

I'd mastered the grid and become a speed walker. I had publishing contacts and I was officially representing Holographic Communications Corporation on the West Coast. I had fallen in love in more ways than one.

CHAPTER 9

DEER IN THE HEADLIGHTS!

Arriving back in L.A. after an exciting month in Manhattan, I was greeted by a patchwork of emotional landmines. I had one foot in Hollywood and the other one in holography, and they seemed to be constantly shifting, even moving in opposite directions. My sanity was being stretched to its outer limits.

I was so focused on the moment to moment, on getting from one place to the next, and on meeting a catalyst, whomever that might be, that I ignored the warnings.

Holograms had become an addiction. Showing them, talking about them, taking orders for them, being photographed wearing them, being called Laser Lady—all of these intoxicants enveloped me. I could barely tolerate sitting at a typewriter for endless hours when I could be out and about with artists and scientists. Little did I know how my pioneering efforts would come back to haunt me.

The Writers Guild was still on strike. My mother hated her Mother's Day scarf because I didn't have Bloomingdale's put it in a box. Selwyn called to say that he and Cecile had had a falling out, and that there was a new exploit with Lloyd Cross.

Richard Harris was filming *The Deadly Trackers* all over Mexico.

Tuesday, May 22, 1973

Meeting—6 PM at Don Richetta's. Bob Gilbert, Robert Curtis, Terry Andrews and I began discussing the Revelation Co. and Fidallgo and the money Fidallgo has put into producing Revelation II. Terry squirmed as Curtis told him he deserves a gold star for monies spent. Nobody asked him (Curtis) and no one asked me. Richetta arrived late. My option money was raised and Richetta asked if I would be willing to take less than $250 per month. As a producer I was supposed to be sympathetic. I flatly said I wouldn't take less. By the end of the meeting I began to think I'd gotten into a group who'd bit off far more than they could chew. I was holding all the cards and they weren't smart enough to recognize it. Richetta said that he didn't want to expend more money on my option. It seemed to me that the only reason I needed them at all was to get the package together, and if they didn't have the balls or the talent to get the money to have a budget drawn up, etc. they shouldn't be involved.

When Walter Lorimer drew up the *Revelation II* option agreement, signed on February 7, 1973 by Robert Gilbert and myself, I was a partner in the Revelation Company. Or at least that was what I had been told.

What I didn't know was that the Revelation Company needed funds to meet its basic requirements—telephone, travel, legal fees, and a demonstration hologram—about $2,500. Gilbert and Richetta linked arms with Andrews and Commons, each assigning 30 percent of their interest in the Revelation Company to a new company called Fidallgo. This was all done in secret. Robert Curtis, who was supposed to direct *Revelation II*, supposedly owned 30 percent of the Revelation Co., and I supposedly owned 10 percent. No one had ever mentioned being a subsidiary of another company to either of us. I felt foolish for being so naïve, and physically ill at the thought of the betrayal.

Meanwhile, they thought by folding me into the production team, I'd leave my writer's hat at the door. That was their game plan. Not mine. And, due to the seemingly endless Writers Guild strike, they had never become a signatory, and since I was a member it meant that my union had no jurisdiction over them. I was relegated to the world of "damned if you do, damned if you don't."

The following day I started taking Valium. My whole world, a world intact one month ago, was now crumbling around me. I went to see my producer friend, Sheldon Davis, who advised me to end the option agreement. "Write a letter of resignation and end the relationship. You can make a much better deal," he assured me.

On May 24, 1973, I sent a letter of resignation to each member of the Revelation Company –Robert Gilbert, Don Richetta, Robert Curtis, and myself—and also sent a copy to Walter Lorimer at Loeb and Loeb. This is important because of the Fidallgo situation, over which I had no control. In my letter I said: "This is to officially notify you that after careful consideration I have decided not to renew the Revelation Company's option on my property *Revelation II*, which expires June 7, 1973."

I went on to say that if they could provide me with copies signed by a major studio or a responsible financier agreeing to commence principle photography by August 7, the company could assume that our original option agreement would be open to re-option. If they did not intend to exercise the option, the letter would serve as my resignation from the Revelation Company. I then got on my father's boat and spent Memorial Day weekend at Catalina Island.

This is what happens when you're young, inexperienced, and emotional. In light of what I know now, it was totally unrealistic to tell them they had to commence principle photography by August 7 or else. I was poking the bear in the eye.

Thursday, May 31, 1973

Met Lynn (Lenau) and Jack (Calmes) at the Hyatt House for breakfast. Went from hyper to calm. Jack set me straight on business, script, holography. I feel 100 percent better.

I was set to veer away from the Revelation Company until Bob Gilbert told me that Walter Lorimer intended to personally present the script to Lew Wasserman at Universal and whomever else it would take. It sounded real.

Lew Wasserman one might say was the most powerful man in Hollywood. As head of MCA, a media conglomerate that merged with Universal Studios to become MCA Universal, if Lew Wasserman gave a project the green light, it went forward.

> Friday, June 1st
>
> Just about to go meet Lynn (Lenau) when Gilbert comes by emotionally carrying a 600-pound weight. He becomes melodramatic wanting to know Why?!? I broke the faith and wrote the nasty letter. I explain that I don't like what happened with Terry Andrews, Dave Commons, and Fidallgo. Never telling (Robert) Curtis and I what they were doing until after the fact. Bob not doing his homework—not knowing how to illuminate the prototype hologram. Going to McDonnell Douglas without telling me, etc. Bob will not admit weakness in any way. I explain that I feel left out, that much is going on behind my back. I want 5 percent of the budget, not 2 percent. I want an agent to look at the script. I'm having a nervous breakdown.

I saw Lynn and Jack and they both told me not to be emotional. If Bob Gilbert and his partners wanted me to give them a two-month extension, fine, but nothing more.

I spoke to Walter Lorimer who said, "If you don't choose to renew the option, we'll get another laser script and we'll continue without you. If you don't feel they're competent, don't renew."

Lorimer's tone and threat were crushing. He was a legal goliath and he was telling me that my script could be tossed in the trash. They'd find another laser script. As if there were dozens of them floating around Hollywood.

"Bob Gilbert spent three weeks in San Francisco and he still doesn't know how to illuminate a—"

"I don't want to hear about it!" Lorimer barked. "I don't care what he did or didn't do!" He hung up.

Well, I guess that knocked the wind out of me. Looking back now, I see that Walter Lorimer had a relationship with Lew Wasserman and MCA. In fact, he had represented MCA. He was firmly on the business side of the table: older, wiser, and clothed with authority. What, he must have asked himself, am I doing with these neophytes? Is this 3-D concept worth the aggravation?

I'm guessing the real reason Lorimer persevered with us was because Sam Gilbert, Bob's father, was one of his most important clients and he didn't have a choice. In my opinion, the mistake the attorney made was going behind people's backs and disrespecting the Writers Guild. At one point he told me to rewrite *Revelation II* in spite of the strike, and if the union expelled me, as long as I paid my dues, they'd have to allow me to work. This was the advice of a studio chief as opposed to a personal attorney giving a young screenwriter sound advice.

In 1973 I had no idea that Sam Gilbert was the shadowy power-broker who would one day tarnish Coach John Wooden and UCLA's stellar basketball history. Known to the players as "Papa Sam," he befriended them and negotiated professional contracts, only taking a dollar for his services. He got them what they wanted for a pittance of the cost and many of the players hung out at Sam Gilbert's house. They loved him.

All I knew was that he was a very successful building contractor who had become the star booster of the UCLA Bruins basketball team. I'd gone to USC, UCLA's arch-rival, so I wasn't that focused on them.

Eventually, Sam Gilbert's passion for basketball victories and disdain for what he considered to be arcane NCAA rules would land him at the center of a serious scandal. In 1981 UCLA was ordered to sever ties with Sam Gilbert.

This is important for several reasons. Even though Bob Gilbert had created magical outdoor sculptures and won an Oscar, he was competing with star athletes for his father's attention. His marriage had collapsed and he wanted to pioneer something that would wow the world. He had moved in with Don Richetta and his wife and, like me, he was dependent upon a temperamental scientist to convince financiers that three-dimensional Integral holography was ready for prime time.

Lloyd Cross was two months late delivering the prototype of the Christ hologram, and Walter Lorimer was not about to present anything to anyone until he knew it would work on a large scale.

On June 4 we convened at Don Richetta's house at 8 pm to sort everything out. Don's wife Debbie began by telling me that I owned at least 50 percent of the property. I took her statement to mean my original screenplay, of which I owned 100 percent. She confided that Walter Lorimer was as fed up with Bob Gilbert as the rest of us. She suggested I hang up on him if he continued to evoke an emotional response. She polished off her advice saying, "He doesn't deal with anything in a professional way."

She knew of what she spoke since Bob had been living there, and now, according to Debbie, was out. "He's full of pipe dreams," she lamented. "Lew Wasserman won't be a reality for a long time. They need a complete presentation."

Once again Lloyd Cross and the School of Holography had become a stumbling block. We were told that Lloyd had been paid to make a hologram for Universal in 1970, and then he disappeared. Once found, he'd been pressured into completing the assignment.

It was now three years later, and it didn't seem as if anything had changed. In light of Lloyd's behavior, Bob Gilbert, Don Richetta, Terry Andrews, and Dave Commons instructed Lorimer to draw up a contract to bind him, not to allow him to make any life-sized holograms of Christ unless they were for the Revelation Company. Essentially the contract would tie Lloyd up, but the way Lorimer wrote it, it gave the Revelation Company the freedom to use other holographers, a convenient loophole in case Cross didn't work out. Bob Gilbert took the contract to San Francisco, agreed to a couple changes, and Lloyd signed it. Lorimer, Richetta, Andrews, and Commons were not happy with the changes.

Terry Andrews and Dave Commons—where did they come from and how did they become the central figures in Fidallgo, a company that had swallowed up the Revelation Company? I'm guessing that they had a connection to Sam Gilbert and Walter Lorimer. Don Richetta seemed to be their link. They had capital to invest when no one else did.

At the May 22 meeting, Terry Andrews warned that if the Christ hologram didn't arrive within the week Fidallgo wouldn't put up any more money. Richetta seconded the threat. I asked how much had been expended and no answer was forthcoming. Suddenly, I realized that the hologram was more important to this group than my screenplay.

When the multiplex hologram finally arrived from San Francisco, I called Charlie Patton, who rushed over with a small laser. Hard as we tried we couldn't find the image.

These were early days and the first integral or multiplex holograms could only be reconstructed using a laser.

The Christ model had been shot on a slowly rotating platform with 16 mm motion picture film. Approximately 1,080 frames had been processed and put into Lloyd's optical holographic camera. Each frame had been projected and recorded onto holographic film, twice. Then mounted onto a cylindrical Plexiglass frame with a motorized base. In order to see the 3-D image floating in space, a laser had to be positioned precisely. If one element was off there was nothing to see.

I arranged for us to meet at Coherent Radiation, where we could illuminate the hologram with one of their more powerful lasers. Terry Andrews and Dave Commons swooped in, lurking, waiting to see the fruits of their investment.

"I don't see anything. Do you? Can you see the image?"

Loud whispers started bouncing off the darkened walls. Low hysterical voices imploring the image of Christ to appear. After about half an hour of watching Dave Schmidt, one of Lloyd's most trusted San Francisco coworkers, and his assistant try to raise the image, they realized that a lens had been left off the viewing stand in San Francisco. That was also why we couldn't see anything with the first laser. Without that lens the moment of triumph was a titanic disappointment.

Terry Andrews and Dave Commons walked out in disgust. Ultimately, the lens was sent along with Lloyd's presentation notes, but the damage had been done. Lloyd Cross' credibility was shot.

—

With all of the technical craziness going on around me, not to mention the intrigue surrounding the Revelation Company, Fidallgo, and my so-called partners, I knew that there was only one person who could unravel the ball of mysteries—my astrologer.

On Thursday, June 7 I had an astrological reading with William Royere, a diminutive man with deep, dark circles framing brown, Sicilian-American, all-knowing eyes. He was a gifted astrologer with an intuitive sixth sense who attracted everyone from Wall Street types to rock 'n' roll royalty to his small tract house in the San Fernando Valley. When he spoke to me in his nasal Brooklyn intonation, I hung on every word.

William told me that if I did not stick with the current Revelation Company, I'd lose everything to do with my screenplay. It was time to revamp. Tie loose ends. He said there was a secret negative energy. It pointed to Don Richetta. He said Robert Curtis was already out of the picture. Bob Gilbert would be put in the right channel and Lorimer would pull everything together.

William also predicted that I'd hear from Richard Harris in about sixty days. I didn't quite understand what he meant, but that didn't matter. Things should start popping in late July and early August. It would be best to make the screenplay deal then.

> Wednesday, June 6th
>
> Met Jerry Gordon, Greg Landini and concert promoter Boyd Grafmyre at the Beverly Wilshire—talked holograms in relation to rock concerts. Turned them on to Jack (Showco) and Selwyn and Cecile (HCCA). Feel good with them—they're an up!

I spoke to Cecile and noted that Jerry Gordon would be in New York the following week to consider making a deal with HCCA. It seemed like everyone was low on funds, but William said it would start flowing in July.

> Friday, June 8, 1973
>
> I went to Loeb and Loeb for an 11 am meeting and found Terry Andrews there. He said he had business on "The Wooden Project" and wanted to hear what I had to say. Walter Lorimer gave me the option agreement. There was a clause x'd out and an amendment that gave all rights to licenses and merchandising to the Revelation Company. Lorimer told me to go ahead and show it to my attorney.

When I left Lorimer's office I felt good, as if things were moving forward. I had given the Revelation Company a two-month option extension. It wasn't until later, when I had been referred to a lawyer of my own, that I realized that I'd met the wolf in sheep's clothing head on. My new lawyer was Andy Pfeffer at Mitchell Silberberg & Knupp, a law firm even larger and probably more prominent than Loeb and Loeb. He asked for all contracts, letters, and my original screenplay. I felt confident that he'd be able to negotiate with Lorimer and that we'd have a production deal in place by the first week in August.

One of my major concerns centered around Walter Lorimer saying that he didn't know if the Revelation Company would sign the Writers Guild agreement when the strike was settled. He said I could challenge them (the Guild) and win. This was meant to scare me because of my resignation letter. It worked. My emotional quotient was so compromised that I woke up crying, took a Valium, and tried to regroup.

Andy asked if I wanted to go forward with the partners. I said no, but I didn't think I had a choice. *Revelation II* was time-sensitive because so many people wanted to be first with these audience-grabbing 3-D special effects.

> Tuesday, June 12, 1973
>
> Today is supposed to be D-Day. Went to office—waited all day.

> Nothing. Spoke to Andy Pfeffer and learned that my contract was not fair to me. I was distraught. Came home and collapsed. Got zonked on tranquilizers after I spoke to Richetta who read me the riot act, saying that I must be alone and confused. Terry Andrews said he read the contract and felt that it was fair to all concerned. I cried for a couple hours. Daddy tried to comfort me. William told me I must act fast if I'm going to save the project. I feel lost and cornered.

On Wednesday I pulled myself together, representing HCCA by showing their holographic jewelry to a department store buyer for Joseph Magnin. The reception was positive, but to get a purchase order I'd have to contact the head buyer in San Francisco.

I called Crayton Smith, the agent I was hoping would represent me. He liked *Revelation II*. This was huge. I felt like I finally had someone in my corner. He asked if he could peddle my screenplay, and I told him I'd call after I met with my attorney. I've always been mindful of protocol.

At 5 pm I met with Andy Pfeffer, explaining what had happened. He showed me where the contract was weighted in my partners' favor and said that he was going to try to renegotiate with Walter Lorimer. At the same time, we would try to make an alternate deal with another producer. He mentioned Frank Capra, Jr. as a possibility. The race was heating up.

Thursday I xeroxed five copies of the script and ran to Crayton Smith's office on Sunset. I was paranoid and my nerves were shot. I was a wreck because I knew if one of my so-called partners saw me, he'd go berserk. They'd gone ballistic when I told them I'd given the script to Selwyn in New York and he loved it. "How could you do that?!" they said. "He's from New York, he'll steal it!"

Their paranoia made no sense to me. I knew Selwyn was holography savvy and could help find someone other than Lloyd Cross to make a life-sized hologram. He wasn't the problem. They were.

I called Andy Pfeffer, who informed me that I had to take a letter to Lorimer's office stating that they must renegotiate or Lorimer would still have the power to make a deal using the old option regulations. The letter, however, could not say that I was severing my relationship with the Revelation Company. I wrote a draft and took it to Loeb, who got Lorimer's secretary to sign for it. Now, Lorimer couldn't make a deal without renegotiating my contract.

Late Saturday, June 18, I called Robert Curtis, who said he hadn't received anything from The Revelation Company. It was as William, my astrologer, predicted. Curtis was already out. Then Bob Gilbert called, all soulful and mellow, asking why I abrogated my relationship with the Revelation Company. I heard an extension phone lift. "Who's listening, Bob?"

Finally he admitted that it was Don Richetta. We were now into the weeds of conspiracy. I had to convince them that they'd misunderstood the meaning of my letter giving them a two-month extension to make a production deal. Yes, I told them, I understand, sometimes it takes longer. If they were working on a deal and it was going to take a little longer, of course I was onboard.

At 11:30 pm Richard Okon called from New York. He'd just spoken to Dermot, Richard Harris' brother, in London. We spoke briefly about Richard Harris, the only bright spot in my complicated universe.

The following morning I got a call from Pfeffer's secretary. "Andy needs the original Revelation Company partnership papers." I called Walter Lorimer, who told me they had never been drawn up; that the option deal I had signed in February with the Revelation Company was worthless. There was no Revelation Company in February, or March, or even April.

In a state of suspended animation, I was getting ready to go to an important Writers Guild meeting that would decide the fate of the strike when my producer friend Shelly Davis called and insisted I stop by. I hadn't been in the door five minutes when he handed me Friday's *Daily Variety*. On June 22, 1973 in *Variety* under "California Incorporations" it lists:

"The Revelation Corp., L.A. County; (100 shares-no par). Sell films. Goggin, Goggin & Commons. 3700 Wilshire Blvd., L.A. Drts: David B.

Commons, 5549 Green Oak Dr., H'wood: Terrence W. Andrews and D.P. Richetta, 9661 Yokum Dr., Beverly Hills."

For me this was a punch in the gut. "If you don't want to extend the option, I'll find another laser script..." echoed in my head until I had a migraine headache.

Lorimer, who was supposed to be representing the partners in the Revelation Company, had hung me out to dry. He told me the company would become a signatory to the Writers Guild then assured me that it was more favorable for us, due to the strike, to not have signed the Guild agreement. He told me to disregard the strike and rewrite my screenplay for the good of the project, strike or no strike, and without rewrite money.

It is my contention that Walter Lorimer encouraged a hostile work environment with my so-called partners. If he had been dealing with a male screenwriter, I'm quite sure he would have handled everything in a more equitable manner. He represented management well. That was his talent, his strong suit.

I was young, optimistic and eager to make my movie. Having been deceived, I turned the blame inward. How did I allow myself to get into this situation? I beat myself up psychologically. I blamed myself.

After reading the entry in *Variety*, I called Don Richetta and asked about the corporation. He indicated that the papers had been filed a month earlier, and that the project was in the toilet because the Guild strike was going to be over and the Revelation Company was not going to become a signatory. Richetta was nice, but I still had a feeling they were secretly proceeding without me. I went to the Guild meeting and sincerely felt that the group was not going to accept the contract. Hostility built—We've gone THIS far—Let's go all the way! A smart person indicated that a two-thirds supermajority of those present was necessary to postpone a vote. It was not reached. The vote was sudden—spontaneous—70-30 victory.

The strike was over for my purposes. I called Robert Curtis and read him the California Incorporation notice. Now he too was outraged and ready to take action. At 9:50 pm Bob Gilbert called and innocently explained that he didn't know about the incorporation papers and would find out what was going on. I told him Richetta said the papers were filed a month prior. Gilbert was not upset; rather he rationalized the action as a logical means of expediting the project.

"Why then were we not told?"

He didn't know, but continued to defend partners. "Revelation Corporation is a subsidiary of Fidallgo."

Gilbert told me to call my attorney and have him call Lorimer and state my terms. I agreed. Then I called Crayton (my new agent) who said that he would arrange a letter with Andy to take an option on *Revelation II* to make me secure. He said that Lorimer should call my attorney.

Just as I was feeling like I was in a tumble dryer, Richard Harris came to town and cheered me up with his enormous kindness. He had never received my letter in Mexico and now he was on his way to London. He'd be back the 16th to do a film for John Frankenheimer at 20th for eight weeks. But even his mellifluous voice wasn't enough to pull me out of my morass. I spent Fourth of July weekend on my father's boat at Catalina Island trying to get my head together. When I had left for New York everything made sense. Now nothing made sense. I agreed to have one more face-to-face meeting with Walter Lorimer.

I met Bob Gilbert at Loeb and Loeb, expecting Lorimer to make things right. Everything began amicably until I was told that I had to sign over all my rights to licensing and merchandising to the Revelation Corporation. I blinked. A dull thud, a shocking thought hit me. I looked at the contract, a contract for a corporation that I was not legally part of. They wanted me to assign my ancillary rights to their corporation. I was confused. If my name wasn't listed in the corporate filing, how was it that I owned 10 percent of the company? I tried to wrap my brain around Lorimer's explanation and couldn't. I looked at the attorney and at Gilbert and stood up, gathering my belongings.

"If I'm smart enough to write the script, I'm smart enough to get up and walk out of here."

There were voices, but I heard nothing. I picked up my bag and kept walking.

The following day my close friend Jenny Corey was leaving for Eugene, Oregon. She urged me to come with her. "Oregon'll turn your head around. You need to get out of L.A.!" Jenny was right: Oregon would usher in a whole new chapter. A happy, healing chapter—grinding wheat and baking bread, jumping off a cliff into cold, clear lake water, hiking in the woods—and leaving Hollywood in the rear-view mirror.

CHAPTER 10

A HIPPIE IN OREGON

The 1970s were good for one thing: allowing young people to tune in to the world around them. This encompassed everything from connecting with other humans to appreciating the air we breathed and the green grass we sprawled across. It was a time of awakening, and when I reached Eugene, Oregon, and Alice Corey's beautiful antique-filled house, bordered by woods and a babbling brook, I opened my eyes. I was home.

At this moment in time, Jenny's mother, Alice, was traveling in Europe, and the large house was filled with her children and grandchildren. There was something about living in a natural setting that brought out our creativity. Bonnie Corey, Jenny's younger sister, and I started working on a children's book called *Gerald the Curious Flea*. I was the writer and Bonnie, a very talented artist, the illustrator.

The idea for Gerald came to me in a dream. There were three cats that had the run of the house and woods. Combine them with warm weather and there were fleas. I figured, why not create a curious flea so fascinated with the human ear that he jumps inside to investigate, gets trapped by wax, and ultimately is ejected? It was my idea to show the inner workings of the human ear using this curious character. At the time Bonnie and I thought it was wonderful, but in retrospect, a talking Q-tip might have been preferable.

Crayton Smith, my new agent, would call with updates. He liked my writing and was very supportive of my decision to walk away from the Revelation Company. That said, the Revelation Company wasn't happy with my departure. Bob Gilbert kept calling and asking how I could throw everything away. He kept insisting that I sign over my ancillary rights to action figures, t-shirts, holograms, etc. to his shady shell of a company. He took no responsibility for the dishonest conduct that prompted my exit. The good news was that I no longer needed Valium. My mental health was improving.

> Tuesday, July 24, 1973
>
> Maitraya came out in the evening. He teaches yoga and guitar. Very spiritual, glowing person. Pure. Jenny, Maitraya and I discussed our raison d'etre, where we're going and what we hope to accomplish in this life. Learned how to read I Ching coins. Listened to Maitraya sing and play original songs on guitar. Very uplifting evening.

I should interject here that while I was growing up my mother was in the background telling me what to do, what to wear, what to say and what to think—unless I was away at summer camp. I suspected that I would always hear her fretful, worrisome, nagging voice beating discordant rhythms in my brain, but being in Oregon was liberating—freedom reigned, and I could embrace my own thoughts and style. The Coreys encouraged me to be myself, and if I made a mistake, fine—I'd have to learn from it.

> Friday, July 26, 1973
>
> Kids tore wallpaper off Celeste's room and couldn't go to the falls, so Bonnie and I went alone. We climbed to the top and rather than climb down I jumped off. It took me a few minutes to get up my nerve as it was about 25 feet, and there was no edge to stand on. I got a running start and jumped. It seemed like

forever blue water and air. I lost the top of my bathing suit and my seashell pin, but it was a master accomplishment to have jumped that far. Bonnie walked down. I hit so hard that it felt like I broke my jaw, but the pain went away.

After I moved into the house in the hills near Mulholland Drive, the owner got a divorce and had to sell it. I'd been there almost a year when I packed my life into boxes and moved in with friends. I knew what I wanted, and I was willing to live like a gypsy until I got it. At this point I must draw your attention to the house where I grew up in Brentwood. Much as I disliked my parents disdainful attitude towards me, I was still their only child, and as such if I wanted to stay there my bedroom was exactly as I'd left it during high school and college. The house sat on an acre of prime real estate while my bedroom was the size of a large studio apartment. So, when I say I was a gypsy, I was an idealized gypsy. I knew, if need be, I had an antique Victorian bed with clean, pressed sheets waiting for me at 555 North Bristol Avenue.

I also believe that this address colored Walter Lorimer's actions towards me. I'm guessing that he assumed I had a trust fund. Unfortunately, for both of us, he was mistaken.

Sunday, August 5, 1973

Crayton called just as we were leaving for the falls. He read me the nice letter from Richard Zanuck and David Brown and told me to stay here and write. I said, "Come up" and he said he'd never leave if he did. And, it's true—there's so much peace and so much beauty and so much love, it's a true miracle of life. He wants me to send him a few story ideas and make a few changes (to Revelation II). He really likes my work.

I was in a holding mode, showing people holograms, polishing *Revelation II*, visiting Klamath Falls. I spoke to Richard Harris, who was leaving L.A. the following day for Washington, and there I was, right next door in Oregon. This made the possibility of seeing him in Seattle very real. Frighteningly so.

When I first met Richard in New York, I was on top of the world. I felt like a shooting star, a screenwriter on the brink of success, and a voice for a new 3-D process that would change everything from gift wrap to information storage, movies, to x-rays. I was a beacon of light and energy.

Now I had little, if any, confidence or self-esteem. How could I have believed Bob Gilbert, Walter Lorimer, and Don Richetta? Lloyd Cross? What was the matter with me? Was William, my astrologer, right—no one else would produce *Revelation II*?

My light had gone dark and it was only through a very special group of close friends that I was able to pull myself together.

Another actor with whom I had a great friendship was Robert Easton. Tall, with white, white skin, blue eyes, and red hair—I knew Robert's face from the 1965 film *The Loved One*, and a host of TV series from *The Red Skelton Hour* to *Mod Squad*. Robert had begun his career on the radio at age fourteen on *Quiz Kids*. He was brilliant with a comedic flourish for playing the buffoon. Robert and I connected because of our mutual love of books. He and his wife, June, had lived in England, and during that time he'd recorded as many speakers as possible. What began as a way of correcting a childhood stutter had become a passion, pushing him to amass the most comprehensive dialect/speech library in the world.

As the years went on, he put coaching ahead of acting. His clients included Sir Lawrence Olivier playing a Michigan auto executive in *The Betsy*, Gregory Peck speaking with an accurate German accent for *The Boys of Brazil*, and Al Pacino sounding Cuban in *Scarface*. He became known as "The Man of A Thousand Voices."

Every time he went on location, he sought out bookstores and searched for old primers on speech and dialect. His wife wanted him to stop collecting. I, on the other hand, was very encouraging. That was obviously his passion, so why not embrace it. I felt that we really listened to one another and had each other's backs.

August 12, 1973

Robert (Easton) called to say that he got my letter—wants to come up but doesn't think he can get away. He spent a week in NY with John Gavin on Long Island. Robert is a love! Made me feel so good and loved. He's going to start a film with Jason Robards and still go to Sweden. All is going well for him. I told him about my happiness in Oregon and assured him he sent me to the right place. He also related to my extreme depression before leaving and my current manic state.

Of course, my fantasy was that I'd walk into the Sea-Tac Hilton, Richard Harris would see me, and our New York connection would reignite. He would save me. And I would help him stop drinking. Cliché as this is, the thought of being with a creative genius who would allow me my own creative space kept me going. I was running on hope-filled fumes.

The Corey sisters found the prospect of my romantic interlude with Richard cause for celebration. They jumped into stylist mode, pulling clothes and accessories out of their closets and my suitcase to put together outfits that were alluring yet tasteful. They also put me in touch with a woman who had dated their father in high school, and now lived in Bellevue, Washington, a suburb of Seattle. Since I was hanging on by a financial thread, having a contact up there buoyed my confidence.

Wednesday, August 16, 1973

Got dressed to kill and missed the plane. Bonnie kept saying, "Oh, we'll make it, don't worry." We arrived too late for the only afternoon flight or possible connection (to Seattle). I was gravely disappointed. Went to the train. "Oh, if only you'd been here ten minutes ago." And I was near tears. "I'll take the bus."

> Got a ticket, figured I'd saved $15 from the plane, but felt missing the plane was a bad omen—maybe I shouldn't even go. All the self-torture, fear, curiosity, passion, hope...

As you might guess, this was not my plan. I'd spoken to Richard, and in my reverie I told him I'd see him very soon. I decided to surprise him.

It was 1973 and there was no Internet, there were no cell phones, and texting had not yet been invented.

The next morning I was dropped off at the Greyhound bus depot. My bus to Seattle was filled with older, blank-faced people. I was sure some creep would sit next to me, but no, a really attractive young man asked if he could sit down. We started talking immediately, he about Vietnam.

He introduced himself as Brian and proceeded to tell me he'd been declared dead in action because an IED blew up and his company left him. The retrieval company picked him up and saved his life. There are dark, confusing edges to this story.

Brian said his wife was notified of his death, and when he arrived home two years later he discovered that she didn't know he was alive and had had a baby with another man. He was crushed and, at twenty-two, he was starting over.

He had just packed his 1929 Cord into his grandfather's garage and was taking the bus to Seattle to catch a flight to California to reenlist in the Marines. As the beautiful Oregon, then Washington countryside flashed past, Brian and I talked about ideas, yoga, fate, sex, philosophy. I told him about Richard Harris, my hopes and my worries. I had no way of knowing if there would be a happy Hollywood ending.

When we reached Tacoma I had to make a decision: change buses for the Sea-Tac Airport and Richard's hotel, or go all the way to Seattle with Brian.

"You know, I'm leaving early tomorrow morning. You can have my apartment. It's paid for til the end of the month. That way, if things don't turn out the way you want, you'll have a place to stay."

It was the end of the day and I was a nervous wreck. I needed time to compose myself, and Brian seemed as genuine as anyone I'd ever met.

He said I could have the bedroom and he would sleep on the couch, and I could have everything in the apartment because he wasn't coming back. I weighed the pros and cons and decided Brian was a gentleman, not a serial killer. I accepted his generous offer.

We ate powdered scrambled eggs with spam for dinner and I had English breakfast tea. At 7 am I heard the front door close. Brian was gone.

The apartment was in a government-owned building that housed a battery of the elderly and alcoholics. It was clean with the bare minimum: a couch, a queen-sized bed, an iron and an ironing board, a couple pots and pans, a few dishes.

Friday morning I called Dorothy, the Coreys' friend and discovered that she had two star-struck teenage daughters. She picked me up and took me to lunch whereupon I explained my reason for being in Seattle.

"You know, Richard Harris is my daughters' favorite! They have all his albums."

"Dorothy, would you like to go down to the Sea-Tac Hilton tonight? I can't promise anything. I don't know where his head will be."

And the die was cast.

Her hair was grey, but the beauty of her youth radiated along with her unbridled enthusiasm. For her this was a very exciting break from her usual routine. She didn't like the location of Brian's apartment and suggested I stay at her house in Bellevue. "Safer," she advised.

Dorothy wanted to introduce me to a young couple from California for whom she babysat. Tom Lynott was the handsome young president of Chris-Craft Industries, a luxury American boat manufacturer. "You have to meet JoEllen," Dorothy insisted. "I know you two will hit it off."

Hit it off we did. JoEllen Lynott looked like Natalie Wood in *Splendor in the Grass*. She was beautiful, as was her son Beau, who was about to turn three. Tom and JoEllen had designed and built a 4,000-square-foot modern, two-story home that backed up to woods, and as luck would have it, they loved their babysitter, and their babysitter loved them.

Dorothy went home to break the exciting news to her daughters while JoEllen let me use her hot rollers. She was a California girl who obviously missed her sunny home state. Dorothy and her bubbly daughter Debbie picked me up at about 8 pm, and off we drove to the Hilton. I knew nothing about the layout of Seattle or Bellevue, so even if I had had a car I would have found the drive a challenge without today's GPS. I was grateful to Dorothy for being so supportive and eager.

> Friday, August 17, 1973
>
> I found Richard in the coffee shop with a beautiful young woman who I would soon learn was Ann Turkel, his co-star. The first thing he said to me was, "You look great!"
>
> He invited me to join them, but I was so nervous my heart was thumping into his dinner. I shakily extracted the pictures of Alice's house and Cottage Grove. Richard told me to wait as they had to see rushes. I could see that he was smitten with Ann. She was lovely. I ate a shrimp cocktail and they came back. I explained about Revelation. Richard said, "You know, I've been on a few binges since I saw you." I invite them to the Lynotts' and Richard says, "I'll call you Sunday." He turns around and in his floor length beige suede coat says again, "I'll call you Sunday."

Sunday rolled around and Tom and JoEllen told me to invite Richard out on their boat. It was a large Chris-Craft model, really more of a yacht. "I'd really like to," the actor lamented, "but 20th flew in some production people. We're over budget and I've got to stay here. I'll call you tonight for sure!"

"No, I won't be here. I'll be on the Lynotts' boat."

Life was so different before cell phones.

I moved to Dorothy's house, giving her nineteen-year-old daughter Pam, who was about to move out, anything she wanted from Brian's apartment. Dorothy and her daughters had eleven dogs: a standard poodle and a miniature poodle in Dorothy's bedroom, two Afghans in Pam's bedroom, a pug in the bathroom, two collies in Deb's domicile, two Afghans and a Keeshond in the garage. Bizarre. Debbie wouldn't allow anyone in her bedroom, not even her mother. Her old collie protected her room like a vicious watchdog.

I may have been staying at Dorothy's but I was spending my days at the Lynotts'. JoEllen thought the Ysan dragon hologram ring was fabulous and that there was a market for holographic jewelry in Seattle. She wanted me to meet the jewelry buyer at Nordstrom and at several of her favorite boutiques. Tom liked the idea of using holographic diffraction grating foil on Chris-Craft boats. When the sun hit the material it would dazzle with kinetic flare.

Infinity glasses were also a hit. So, even though things weren't going as I'd hoped with Richard, at least I was doing something productive with holograms.

I saw Richard several more times, but it was never the right time. The film was putting him in a less than stellar mood, and I could see that he and Ann had a special connection.

JoEllen wanted to meet him and I wanted to see him before the film wrapped. We fortified ourselves with wine and made our way to the Hilton. I made the introduction and waited to speak to Richard alone.

"I feel like an ass," I told him.

"Oh, don't feel like that! I'm working! Please don't feel like that." He hugged me and disappeared into the night.

Friday JoEllen took me to meet Mi Flick, a beautiful, bubbly twenty-six-year-old Swedish sophisticate with blond hair and a mysterious hearty laugh who owned a luxurious and hip Seattle boutique. Her soon-to-be ex-husband Bob, formerly a member of the Brothers Four, was a big star in Japan, a country that thrived on science and art. Both Flicks saw holograms as the wave of the future.

Mi immediately gave me an order for ten Ysan dragon hologram rings. It was a start. HCCA would see a glimmer of West Coast interest. I also had an appointment with two jewelry distributors who liked the hologram ring but said it was too special for them. Substitute novel for special and you get their point.

Monday, JoEllen took me to Nordstrom where I got another order, this time from the jewelry buyer. Even though things didn't turn out the way I wanted, I had made wonderful new friends and important business connections.

> Tuesday, August 28, 1973
>
> Hurriedly caught the train. Everyone wanted me to stay, but I couldn't. Train far better than the bus. Can move around comfortably and passengers keep to themselves. Wrote to Richard all the way down. Read Don Juan and arrived to see Jenny and Dave. Seattle now seems like a fantasy, a figment of my ripe imagination. Spoke to mother who was more than friendly since she hadn't heard from me in three weeks.

Timing is everything. *99 and 44/100 Percent Dead* bombed. Richard Harris fell in love with Ann Turkel and they were married the following year. Alcoholism aside, Richard set a very high bar, giving me plenty to think about on that train ride back to Oregon.

CHAPTER 11

PARADISE FOUND

It was now September and I was still enjoying Alice Corey's hospitality in Oregon. I was exploring the great outdoors, camping at Winberry Creek with Bonnie Corey and her husband Jeff Molatore, and writing. Perhaps, for the first time in my life, I was living in the present instead of the past or the future.

Along with the Coreys I delved into ESP, auras, angels, astral projection and lifting each other up using only two fingers, grinding wheat for homemade bread, and making soybean steaks from scratch. We meditated, grew vegetables, and picked concord grapes in an orchard. My time in Cottage Grove put things in perspective. I didn't want to leave, but people in L.A. kept telling me I needed to be there if I wanted to be a screenwriter.

And they were right, but at that moment I was free as a bird. The owner of the house I'd been living in off Mulholland Drive had had to sell it, and, in turn, I had to box up my belongings and store them at my parent's house in Brentwood.

I wanted to live in both places – L.A. would be the business hub, Cottage Grove my creative refuge. Every day I buried my fears, telling myself that hard work equaled success, and when I sold a screenplay I'd be able to buy property wherever I wanted. I gave myself silent pep talks to make a decision. Stay a little longer or go home, or maybe stay forever. "Your life is in boxes, you can do anything," kept echoing through my head. I knew, if need be, I had an antique Victorian bed with pressed sheets waiting for me at 555 North Bristol Avenue.

Monday, October 1, 1973

Finished the final draft of Revelation II. It really is good. Helped Jeff grind wheat for bread. It has to rise three times. I feel like a pioneer. I really am learning how to live. Gaggle of geese in the Row River. Can hear them night and day. Feel time passing and know that paradise cannot go on forever. Need time to think about facing the clangorous music. This has been a perfect environment. I hope the script brings enough money to get my own place. A fresh start.

The coffers were nearly empty and my agent as well as friends and parents, were urging me to reenter the world of the gainfully employed. I had successfully escaped from L.A. and regained a sense of calm. I sensed that if I wanted *Revelation II* to be produced. I needed to jump back into the Hollywood caldron.

Tuesday, October 2, 1973

Got an order for 1 dozen hologram rings from Kaufman Brothers. (Eugene, OR)

I spoke to Cecile in Manhattan. She told me that Selwyn had his first art holography exhibit in New York. She was still struggling with the business. All the promise and vision in the world was useless unless there was financial backing to implement it, and after our unfortunate meeting with Lloyd Cross and Joe Greenberg, one of her best sources had been compromised.

As Dorothy predicted, JoEllen and I had become very good friends. She and her husband, Tom, and I bonded over holography and its infinite possibilities. Being in its infancy, no one knew how holograms would hold up. They were after all, images shot on special holographic emulsion. What would the quality of the film and imagery be like in twenty years? We'd have to wait to find out.

JoEllen knew how disappointed I was over the dissolution of the never-really-formed Revelation Company, and Richard Harris. Her solution was brilliant: "Come to Hawaii with us, then you can fly back to L.A. from there."

It was a dream come true. I'd always wanted to go to Hawaii and the Lynotts were inviting me as their guest.

Saying goodbye to the Coreys was bittersweet. I loved living with them, however I knew there was no way I could fly to Hawaii and not fly home to L.A. I promised myself that one day I would return to Oregon, and maybe even move there.

Monday, October 22, 1973 Seattle Airport

Had to get up at 7 to get it together. Tom, JoEllen, Brian, Kathy, Beau, and I took off for Hawaii. Waited three hours in Honolulu for connecting flight (to Maui)–finally got to Royal Lahaina and ran, fully clothed into the clear, warm blue water. The first day of heaven. Dined and danced at the hotel.

Thursday, October 25, 1973

Wrote Cecile: company categories: advertising, special effects, film applications, stage applications, commercial applications (which mean anything from toys to decorating) In the midst of my Maui bliss, I was still making notes pertaining to HCCA. Perhaps snorkeling in warm, clear saltwater looking at varieties of yellow and purple finger coral and tiny black striped tropical fish got my creative juices flowing. The wheels of my mind were perpetually in motion even when I was sprawled out on a lounge chair with my eyes closed.

One of my great passions has always been listening to people within earshot. Call it eavesdropping if you like, or call it fuel for the writer's soul. Having coffee one morning at the Royal Lahaina's seaside cafe I overheard a man say, "I used to fly first class until I discovered that all the interesting people were in coach."

At the time I thought this was a wonderful observation. It was idealistic young people who were reshaping the world, and, for the most part, flying coach. We were the ones who had to recover from Vietnam, political unrest, and inequality. If we had something to say, we said it.

> Monday, October 29, 1973
>
> Went snorkeling, had trouble with mask and lost an Acapulco earring. Ocean floor covered with sea urchins, brown rock spiny variety and deep red with tentacles lurking from rocks. Touch one and you're in trouble.

At the end of the day we gathered at the water's edge to have pina coladas and watch three-year-old Beau play in the sand. As I watched the huge, bright pinks and oranges of a Maui sunset, I imagined being able to shoot holograms onto clear film that would be adhered to a sliding glass door. If you wanted to see a Maui sunset, all you'd have to do was program that setting the same way you changed a TV channel. (Of course, with today's technology all you'd have to do is say, "Alexa, show me a Maui sunset.") I wrote the following description in my diary:

> Holographic Modular Environs! Modular dwelling with 4'x8' hologram window with 1,000's of settings. Can transform small module into vast scenic vista. Watch Maui sunset turn to night. Programmed like integral hologram. Movement. Can also use as an educational device.

In 1973, it took three hours to drive from Lahaina to Hana on a road wide enough for one and one-half cars. It was well worth the journey when you entered the lush, tropical paradise with spectacular waterfalls, jet black granular sand, guavas hanging over the road, mangoes and bananas for the taking. But there's always a price to pay for this kind of magnificence, and in this case it was tiny biting mosquitos.

The town of Hana was so small there was only one hotel, and we were too late for lunch and too early for dinner. Oh, well, who cares? Grab a guava.

On Thursday, November 1, I had to fly back to Los Angeles. It wasn't easy breaking away from a group of people who had given me a new lease on life. I felt energized, and really, really grateful.

> Friday, November 2, 1973 Los Angeles
>
> Didn't want to talk to too many people. Reclusive. Called Lynn (Weston) and Shelly (Davis)—both in dumps. Charlie (Patton) A-Okay! Stopped drinking heavily, taking two holography courses— UCLA extension and Barnsdall Park. Optimistic! Nice to be back, not bad at all.

I arrived back in L.A. with a revised screenplay, part of a children's book, invoices for holographic jewelry sold in Washington and Oregon, and a deep golden Hawaiian tan. I was primed for success.

CHAPTER 12

BACK TO BUSINESS

Sunday, November 4, 1973

Met Star Rainbow at Topanga swap meet. He's putting together holography people for London play: "Alice Through the Juke Box"—it's what dude on Maui mentioned. Incredible coincidence since he (Star Rainbow) lives in Laguna Beach and creates airbrush clothes.

I was thrilled when Mi called to say she was coming to Los Angeles for a week. She needed sunshine and trendy new clothes for her boutique. She needed a friend in L.A. and I needed to be reminded of the happy times I enjoyed with her and JoEllen in Seattle. Her visit would be a welcome diversion.

Wednesday, November 7, 1973

Sent JoEllen (Lynott) battery of material on holography. Spent two hours on the phone with Richard Harris' friend Richard Okon and then Cecile (Ruchin). Went to Charlie's (Patton) holography class at UCLA. Got all excited! Went to Tana's and played Hollywood, saw old friends.

Once back in Los Angeles I sublet a room from friends in Malibu and used the top of my father's office building as my writing sanctuary and holography base. There was a large empty closet where I could store my treasures: a large unframed painting Channing Peake had given me, a signed copy of Françoise Gilot's book *Life with Picasso*, and the smaller treasures Richard Harris had bought me in New York. I had most of my possessions in boxes since the place where I was staying had no space for storage. Eventually the closet would hold all of HCCA's rolls of small McDonnell Douglas Sun Ra holograms.

My father's personal office was large enough for two mahogany desks, a perfect upstairs hideaway. I had everything I needed: a spacious workspace, plenty of storage space, a telephone, and a typewriter.

One morning Shelly Davis, the producer with whom I'd been developing film projects, called to tell me he'd come up with a great idea for a "movie of the week" and I needed to be at MGM at 2:30 pm sharp.

The two Shellys—Shelly Davis and Shelly Brodsky—had an office and a first-look deal at MGM. First look meant that their production company, Now Productions, would show motion picture and TV series proposals to MGM executives before taking them anywhere else. If the studio liked the idea, they'd make a development deal and money would be forthcoming. If not, the project would be presented directly to a network, or stars would be approached to put a package together.

By now I'd known Shelly Davis for over five years. He was an idea man who had been one of the founders of the famed Whiskey a Go Go on the Sunset Strip. Over the years he'd collected friends from high society, entertainment, music, and politics. He loved to cook and entertain, and his parties were always eclectic, delicious networking opportunities. Rona Barrett, the Hollywood gossip columnist of the day, was one of his closest friends.

Shelly Brodsky was old-school East Coast business. He'd been a talent agent at William Morris and was still managing Soupy Sales.

Davis was tall, obese, and gregarious, while Brodsky was tall, neither fat nor thin, and always New York-professional. They made a good team, but they needed a pot of gold to push projects forward. Shelly Davis' old easy-money Whiskey days were long gone. His livelihood depended upon making deals, and Hollywood deals were fraught with booby traps.

At 2:30 pm I turned up at MGM to find Rene Bond, a young woman with an angelic face and a Playmate's body, her torso and legs squeezed into tight pink pants, and her inflated breasts overflowing a tube top.

She was a porno queen with some 200 adult films under her belt, and a thumb's up from Hugh Hefner. She was on a mission to transcend her bawdy image and become a mainstream movie star.

Three-and-a-half hours later we finally agreed on a story idea.

I enjoyed working with the two Shellys, and in turn they gave me MGM office space. I was their head writer and anything I wrote with them gave me a one-third stake in the project. Of course, for the privilege of being part of the team, when I was there, sitting at the front desk, I had to pretend that I was their secretary. My goal was to become a writer-producer, and I believed that Davis and Brodsky had the studio and network connections to open those doors. Shelly Davis thought the world of porn would be ratings gold. It was, after all, one of America's guilty pleasures, and Rene Bond looking like the girl next door would resonate with puritanically conflicted imaginations. The question was, was network television ready to explore forbidden fruit?

It was during this period that I began spending a lot of time with Shelly's close friends, the Westons. Ray was the Beverly Hills doctor to the stars, and his wife Lynn was an accomplished artist whose work hung in museums. Her latest passion was film. She wanted to become a movie director.

The Westons were my parents' age but they were nothing like them. They inhaled movies and literature, and if you got a sinus infection they could get you an immediate appointment with a wait-listed Beverly Hills specialist. They also loved hosting up-and-coming talent. I enjoyed bringing my friends to their comfortable, welcoming, art-filled Beverly Hills home. The salon setting usually brought older, very established actors and directors together with a new crop of talent—the perfect combination for group inspiration. One of my favorite evenings was spent with iconic writer Anaïs Nin and her partner Rupert Pole. They were both warm and encouraging when I shared my futuristic projections for the visual arts and showed them several holograms from the School of Holography.

Wednesday, November 14, 1973

Picked Mi Flick up at the airport and went to Tana's for drinks —reunion.

Showing Mi Flick Los Angeles was Disneyland-exciting. When she arrived she was completely stressed. "I don't know how I'm going to take care of my son and run the boutique." Divorce was hard and she needed to regroup. She needed a plan.

I took her to the garment district in downtown L.A. to ferret out hot new designers, to the commissary at MGM for lunch, and to gallery openings. It was non-stop, and over the course of a week we became trusted friends.

Saturday, November 17, 1973

Took Mi to meet Lynn Weston. Ray already in bed so the three of us chatted. Mi told of her alcoholic mother with lover and father knowing one another—friends. Pills, liquor—hospital—feigned recovery—home—alcoholic again. All through childhood she needed to be strong!

Sharing intimate details about her life was cathartic for Mi. She realized how much her well-being depended upon sunshine and positive people. La La Land suited her. She loved L.A. and L.A. loved her back.

Sunday, November 18, 1973

Slide/video show at 707. Drank champagne and saw Topanga crowd. Dean (Stockwell) said, "You look glowing!" Funny, I felt that way. Beautiful, hip people. Good slides. Charlie (Patton) drunk. Rainbow (club) early—great seats. Lee Housekeeper joined us with Bob, guitar player scouting work.

The worlds of fine art and holography had begun to collide and mesh, and the first laser light show, *Laserium*, was about to open at the Griffith Observatory.

Monday, November 19, 1973

Laserium show at the Planetarium. Interesting because it's the first, but too long and boring. Crayton (Smith) not too impressed. He said money in Hollywood is hidden until the first of the year. Studios don't know who'll stay in business. Wait for interest rates to lower. Patience. Lunch Nancy Lee (Andrews) and Lee (Housekeeper).

Tuesday would be Mi Flick's last full California day, so naturally she had to have an astrological reading with William. When I picked her up she radiated optimism. The stars foretold success without financial worry. She was greatly relieved.

I had become good friends with a Canadian musician named John Finley from a group called Rhinoceros. He and his musician friend Henry Marx wanted to set Mi up with their friend Artie Wayne, who happened to be a manager and the director of creative services for Warner Brothers Music. Blind date. Why not? Fine, I'd go with her, make sure everything was kosher, then leave.

Tuesday, November 20, 1973

Artie's place "perfect" like a New York garden apartment. His phone rang—business. Talked business and holography. Mi couldn't believe it—he had to meet someone at ten for business so he took us to the Old World for a quick dinner. Artie's a contemporary Sammy Glick, always running and mushrooming projects. Said he'd like to see some holograms. When? How's tomorrow at 2:25? Fine!

The following day Mi and I went to Warner Bros. Records. The following is an excerpt from a letter I wrote Cecile at HCCA:

Artie Wayne, a songwriter and publishing exec. at Warner Brothers Records flipped over holograms. Gave a presentation there Wednesday and all record execs flipped. Artie wants me to set him up with a few holograms (on loan) because they're going to do a big publicity campaign with new offices, and he's assured me that we'll receive appropriate credit. He's agreed to sell them (holograms) in beginning and ultimately would like to work into serious involvement with company (HCCA). He's black, aggressive and could sell the Brooklyn bridge if necessary. The traffic through his office is everyone from Dylan to Lennon and on and on. It's a natural! We'll get national coverage and I'm sure a lot of orders.

Artie Wayne saw the light. It was a very exciting meeting and I felt like I was finally getting someplace. A hologram on a record cover would grab people. Seeing a singer or a group floating inside a cylinder would be marketing magic, and Lloyd Cross was not the only holographer capable of delivering high-quality 3-D imagery.

When we arrived at LAX, Mi turned to me and said, "Linda, if I ever send someone to you, you have to take him seriously. I mean it. I know people you need to meet."

On that high note I dropped her off at the airport and joined friends who told me Francis Ford Coppola would be filming *The Godfather Part II* in downtown L.A., a night shoot. "Tell him about *Revelation II*," they insisted. "Show him a hologram! He taught you to write a screenplay. Make him proud!"

I remember this night well. My friends wanted me to finish what I'd started with Coppola. Let him know that I'd learned something meaningful in his UCLA screenwriting class, something I'd put to good use. If anyone could help me produce *Revelation II,* it was him. That evening was cold in downtown L.A., and Coppola was getting ready to blow up a vintage car. Someone pointed to an authentic $8,000 period lamppost that had been installed for the next shot. There was a mixture of tension and excitement as we all awaited the arrival of two of the director's young children. It was important to him to have them watch this dramatic scene. There could only be one take.

I took that window of opportunity to say hello and show him a few holograms. In the same letter to Cecile Ruchin dated November 23, 1973, I wrote the following:

> Wednesday night I was invited to watch them film a location shot for The Godfather II. I haven't seen Francis Coppola in a few years and was warmly received. I told him about Revelation II and asked him if he was familiar with holography. He said yes, then I showed him a few (holograms) and he flipped! Showed them to everyone on the set. He has the script (Revelation II) and he could make the immediate difference as Paramount is the richest studio in town!

Following this entry, I went on to request stationery and brochures. I told Cecile the brochure should state:

> THEATRICAL HOLOGRAPHY—special effects for live performances, film, or tape. Animation. Laser artistry. Advertising. Create a total environment with the magic of holography and laser beams. I believed that something along these lines was needed to generate interest and start a conversation with film industry executives.

> Monday, November 26, 1973
>
> Nick Franzosa came to the office to buy a hologram and talk business. He's putting up a holographic environment for Host International in Houston. My idea two years ago. It seems Charlie (Patton) turned him on to TRW, McDonnell Douglas, etc. I ended up thinking that I was competing with this man who's way out in front in advertising and special effects. Charlie's a child telling precious, hard learned secrets to anyone who'll listen.

Meanwhile, back at MGM there was interest in Shelly Davis' porno-based movie of the week, and I was tasked with writing a treatment. For research purposes I was sent to a beautiful plantation house with white columns in the heart of ultra-conservative, old-monied Pasadena where the couple who owned it, Talie and Patrick, were using it as a porno set.

Talie was actually a Southern belle who had become an agent for porno players. Patrick, her husband, was a classically trained actor and the film's producer. I was told that they were very discreet—the neighbors had no idea that inside this lovely abode furnished with Chippendale furniture and blackout drapes, cameras were rolling and thighs were slapping, especially in the middle of the night.

What I found most insightful was that many crewmembers were regularly employed by studios and networks. This was a moonlighting gig, and unlike the sleazy, stereotypical notion of dirty porn, from what I observed, this was pretty much like any movie set until the period costumes flew off. It was strictly business and somebody was making a ton of money.

> Tuesday, December 4, 1973
>
> Turned in (porno) treatment. Went to Rene Bond's for dinner. Was wiped out, but rallied. Tony (Rene's boyfriend) is crazy. He spent the evening talking about his record collection—over 300,000 LPs and singles—which burned in the Malibu fire. He made Japanese lumber deal, discovered Rod Stewart and Faces. How he's done all this and come out a reactionary loser is strikingly sad, if even true.

You have an idea, you develop it, you research it, you write it, you ask trusted friends to read it, and finally, you pitch it. Now you wait, and believe me, waiting is the hardest.

This had been a pivotal year not just for me, but for America as a whole. In March our troops were brought home from Vietnam, and Richard Nixon was starting his second term alongside the growing Watergate scandal. Former president Lyndon Johnson was dead, and Vice President Spiro Agnew was forced to resign because of tax fraud. There was an energy crisis closing in, and predictions of a February blackout loomed. In spite of all this, Hollywood was still doing business as usual.

> Thursday, December 6, 1973
>
> Gone through so many changes, don't know who to trust. Spoke to Cecile (Ruchin) who told me everything is taking shape in New York. She'll call me Tuesday. Just about to leave when Eddie Auswachs called and made an appointment for tomorrow.

Eddie Auswachs was a force of rock 'n' roll nature. In his twenties, tall with a white man's curly red afro, he was producing *Alice Through the Juke Box* in London and he needed special effects that included laser effects and holograms. He was in Los Angeles to find state-of-the-art holographic technology, and now that I was working with HCCA, the circle of people agenting for genius physicists was widening.

I met Eddie in Beverly Hills at a house that looked like a spread in *Architectural Digest.* On this warm, sunny December day, his business associates and their French girlfriends were swimming and tanning *au natural* by a pool guarded by lion heads spitting water. It was clear to me that Eddie had serious financial backing.

Afterwards I dropped off a copy of *Revelation II* for Artie Wayne at Warner Bros. Music and drove to Ardison Phillips' studio near downtown L.A. I was in the process of telling him about my meeting with Eddie Auswachs when I "tripped" on mylar reflections. Yes, reflections. It looked like there were sheets of mylar in my path when in fact it was nothing more than a projection of sheets of mylar.

Ardison showed me two parabolic mirrors that were projecting the shiny metal. It was essentially the same effect as a hologram only without special film and a high-powered laser. Strategically positioned mirrors could project what appeared to be a solid image on stage for a fraction of the cost. If Eddie Auswachs could use parabolic mirrors to create the effects we'd discussed he could save $20,000 to $25,000. I couldn't wait to add this to my list of possibilities, and if Ardison could do it that would be a bonus.

My life was an intensely exciting whirlwind, and no matter what I did I couldn't bring myself to step back and take a deep breath. It was as if I'd climbed aboard an amusement park ride and I just had to hang on until it stopped.

> Wednesday, December 12, 1973
>
> Went downtown with Lynn (Weston). In ethnic grocery when well-dressed Chinese speaking gentleman came in. It turns out that he's Hawaiian. Lynn says, "Say hello to my friend Don Ho." He says, "I know him well. I will." To which Lynn asks, "Do you know Ed Brown?" and a conversation develops. Lynn mentions my hologram ring and Leighton identifies himself. He's an interior decorator from Hawaii. He flips over my ring, says he was at Neiman Marcus yesterday and saw a hologram in the photography department. He wants the Hawaiian franchise for Holographic Communications Corp. (HCCA) Make cheap souvenirs. It was a miraculous encounter.

I was feeling frustrated and somewhat defeated when William, my astrologer, called and invited me to his house in the valley for one of his parties. He and his wife Aya always put amazing, high energy people together, and this was no exception.

There I met Tom Wilkes, who was well known for having been the original art director of the Monterey Pop Festival, the Concert for Bangladesh, and had just won a Grammy for best recording package for *The Who's Tommy*. There was a Topanga artist named Cameron who was known as "the woman of the woods," and my old Leon Russell photographer friend Andee Cohen. The house was vibrating with artists, writers, musicians and designers, and I was beginning to feel the same sense of happiness that I'd felt in Oregon with the Coreys. William took me aside and told me I had four to five years to run, run, run. That *Revelation II* would be produced and that I would have my own production company. It's all going to work, he assured me, and by the time I left I was floating on a cloud of optimism.

On Thursday, December 20 I drove up to the Santa Ynez Valley for a few days. This time I stayed with my friend Susan Turnbull at her mother's ranch. She'd been critically injured by a drunk driver on the San Marcos Pass, air lifted to the hospital in Santa Barbara, and miraculously survived.

I had lunch with Channing Peake and his twenty-something hottie from New York who called everyone darling and kept asking me where she should have her hair cut in L.A. The good news was that Channing wanted to introduce me to a man named Peter Burtness, who was a real estate developer and part owner of the Santa Ynez mint. He thought Peter wanted to do something humanitarian and might be interested in investing in holography for educational purposes. Since it was possible to store the entire *Encyclopedia Britannica* onto one 8-inch by-10-inch hologram, holography could be considered a tool capable of delivering information to millions of students.

> Saturday, December 22, 1973, 11 am
>
> Peter Burtness. Showed him integral holography and explained the educational approach. He wants to put out educational films for video system. Can get money from Kroc (McDonald's), Buel, the vet, Hastings Harcourt, etc. Wants prospectus and to go directly into production. Sounds good and will fit right into holography prospects. Build from base up. Left Peter feeling high and mighty anxious to get money for development.

Sunday morning I had another enthusiastic meeting with Peter Burtness before driving two-and-a-half hours to Brentwood to meet my parents before we were to drive down to attend an afternoon Christmas party at my uncle's house in Orange County. After that I was expected in Malibu for dinner at eight.

In the days before freeways became parking lots, it was easy to zoom along the San Marcos Pass, past Santa Barbara to the 101 freeway and Brentwood. From Brentwood you could get on the 405 and in an hour you'd be in Orange County. I can't imagine how many hours it takes today.

Still enthusiastic from my morning meeting, I told my father that I needed to be in Malibu by eight.

"When did you make that plan?!" He was furious.

"Last night. When I was in Santa Ynez."

"You're lying!"

"Why would I lie?!"

The fuse had been lit. My father stormed into the kitchen as my mother said, "Linda, if you don't go I'll stay home."

"How about this?" I countered. "I drive too, and I'll bring mother home early. On my way to Malibu."

"No!" Towering over me and enraged, my father spat, "You're nothing but a leach and a parasite!"

A hot anvil of words tore through me. Crushing. Speechless. I rushed out of the house and drove to Malibu. I'd been away for months, and now because I wasn't doing exactly what he wanted me to do I was an insect. My mother had her own agenda and I was last on the list. The next day my father called to apologize, but the damage had been done. I couldn't unhear his hate-streaked voice.

As I mentioned earlier, my family put the "d" in dysfunctional. My parents had been separated when I was at USC, and now they were making the best of a very unhappy marriage and, as an only child, I was caught in the middle. They wouldn't get a divorce because both of their identities were tied to 555 North Bristol Avenue. I was too entrenched myself to even consider removing the address from my driver's license. Simply as a result of their one-acre property, people respected them, even when they didn't respect each other or, as I sometimes suspected, themselves.

> Christmas Eve 1973
>
> Feeling like dying after daddy's leach-parasite accusation. Channing set up a meeting for me in L.A. with Konrad Kellen to sell a couple pieces of art. Handsome, classical European. Had coffee as my Valium started to take over. He asked me about myself and considered me a gift from Channing. Charming with many ego ups to soothe the pain. He bought two paintings.

Konrad Kellen was a German-born American political scientist, intelligence analyst, author, and friend of Channing Peake. It was Kellen who spearheaded the open letter from RAND to the U.S. government informing them that the communists in North Vietnam had high morale, not low morale, and that the war was unwinnable.

On Christmas Eve I delivered two paintings to Konrad Kellen and received a much appreciated commission from Channing.

Christmas Day was spent dutifully with my parents in Brentwood. To make amends my father gave me $100 instead of his usual $50. He was a complicated man. My childhood knight in shining armor had become bitter and mean towards my mother and in her reflection, me. He didn't mind the idea of my sharing his personal office to write and represent holographers, but he wanted to see results. If I didn't get a show on the air or a film produced, to him that meant I lacked talent. He didn't really share my enthusiasm for holography either. He tolerated it.

My 1973 was wrapping up and I was reflecting on what I would consider to be the most significant year of my life. It was so jam-packed that I broke it down month-by-month at the end of my diary. Here is the recap:

> January was frantic, wondering what the year would bring. Making plans to move to where?
>
> February: Signed option agreement at Loeb and Loeb for Revelation II with Bob Gilbert, met Selwyn Lissack and heard about Dalí holograms, sublet space in Malibu until real money came in, felt on top of the world.
>
> March: moved to Malibu, formed Optical Infinities with Lloyd Cross, wrote notes and took only photos of first integral holography machine. Felt part of history! Made friends with older writers on Writers Guild picket line.
>
> April: Wish came true—got insurance money from Davana Road flood and immediately flew east. New York. Met Cecile Ruchin.

Felt real art world, the money makers, the pros. Fell in love with Richard Harris and became another of his conquests. Got past it cause of business. Charlie (Patton) in New York. Saw Holoconcepts holograms—Cartier hand. Had the time of my life.

May: Back to L.A. met John Finley's friend Henry Marx and got into potential songwriting. Revelation Co., hologram arrived from Lloyd Cross without a critical lens for viewing stand and financiers walked out. Revelation Co. doesn't legally exist, Fidallgo, a new company with different partners is trying to take over. End of writers strike.

June: When Walter Lorimer told me if I didn't want to assign my ancillary rights to a company that had not yet been formed that he'd find another script with a hologram, I fell apart day by day. In one week got Andy Pfeffer (lawyer), Crayton Smith (agent) and hope. Was in a bad mental, financial place. Needed help!

July: Left with Jenny (Corey) for Oregon. William's predictions very accurate. I walked away and was ready for a nervous breakdown. Robert (Easton) told me to go and artist Robert Blue and I played at the Polo Lounge and the next day—wham—off to Oregon. Jenny was right— it would change my life. I would come back to L.A. a different person.

August: Went to Seattle to see Richard Harris—went badly, met Lynotts, and Mi Flick. Became more centered, made a lot of good holography marketing connections

September: Back to Cottage Grove. Again blessed with Alice (Corey) and nature and pure love. Being yourself, not having to be somebody. Enough to live and work and love and be loved.

October: Working with HCCA—Cecile in New York, writing Laser Lady, fabulous trip to Hawaii with Lynotts.

November: Back to L.A., Laserium opens, working at MGM with Shelly Davis. Mi Flick lit up L.A., too many holography marketing connections to list. Asked Francis Coppola to read Revelation II.

December: Desperate to sell project and stabilize life.

CHAPTER 13

RAINBOW BRIDGE

It was finally New Year's Eve and producer Peter MacGregor-Scott and his wife, Nan, were throwing one of their wonderful, salon-inspired parties. I wore a Japanese kimono with a squash blossom necklace. Yes, a Japanese kimono with a silver and turquoise southwestern Navajo necklace. I can't quite wrap my head around that combination, but then we were all young, all ambitious, and all ready to express ourselves artistically.

One of the perks of writing a memoir that includes notable people is that the Internet now provides a roadmap of the last forty years. Sadly, Peter MacGregor-Scott passed away in October 2017 after a taxi accident in New York City. Born in England and extremely cordial and well-liked in Hollywood, he produced dozens of motion pictures, including *The Fugitive*, two Batman films, and three Cheech and Chong movies. And, he knew how to produce a memorable party.

As the clock struck midnight, I welcomed 1974 surrounded by writers, producers, film editors—people working in the industry who boosted my optimism. I had taken a piece of holographic diffraction grating material with an adhesive backing and pressed it onto a bracelet someone had given me for Christmas. Suddenly I had a radiant, kinetic piece of holographic jewelry, and it was a crowd pleaser. I had something tangible. The holographic material could be purchased and molded onto Plexiglass and cut into varying widths. It was a clean, simple, beautiful concept. When light hit it, it would spark kaleidoscopic spectral patterns. Think of the psychedelic seventies.

I wanted to call my West Coast product line "Jupiter 5." Jupiter was my ruling planet and the fifth planet from the sun. It happened to be the largest planet in our solar system. I was now keen on making holographic jewelry.

> Tuesday, January 8, 1974
>
> Saw Charlie (Patton) about Jupiter 5 jewelry. Got drunk on champagne and vodka and could hardly drive home. Alcohol has hit Charlie as it did his mother and there's no visible way to take away the pain. First we argued and next we brainstormed: dye cut fingernails, beauty marks, shoes and on and on. It was amazing—up and down—down and up—played with his new laser making patterns to coincide with mood and beat of music. Kissed Charlie goodnight and for the first time felt his power.

On Thursday, January 10, Lloyd Cross, Pam Brazier, and three others from the School of Holography packed a car in San Francisco, bypassed Los Angeles, and headed directly to Star Rainbow's in Laguna Beach.

Laguna Beach, California, had always been a coastal jewel attracting artists and wealthy vacationers. Then on Christmas day 1970, the stars aligned or misaligned, depending on the side you were on—flower children or conservative John Birchers—and Curtis Reed, who had legally changed his name to Curtis Rainbow, along with L.A. Oracle art director Ted Shields, who had become known as Star Rainbow, promoted a four-day rock festival in Laguna Canyon. Shields managed to get widespread coverage in the L.A. *Free Press*, plus the hippie underground handed out thousands of Orange Sunshine acid tabs on invitations.

Twenty-five thousand flower children made their way to the small seaside town, heading up into the canyon for a concert that got stopped in its tracks by police roadblocks. Not to be dissuaded by the police, a plane flew over, dropping thousands more acid-tinged invitations—showing everyone what it really meant to *drop* acid.

Just as Woodstock became a legendary happening, tales of Laguna went global. Curtis Rainbow opened a vegetarian/vegan restaurant called Love Animals, Don't Eat Them, and founded a hippie commune called Rainbow Island. The local health department made it their mission to close the restaurant, so in 1972 the commune and nonprofit church he started sustained itself by airbrushing psychedelic shirts for rock concerts. The weather was sunny, the air was clean, and life was good.

By 1974 Star Rainbow was running a successful business, and to me he was an imposing alpha-dog father figure. Tall, charming, ultra cool, tan and buff, he directed the Rainbow Kids to airbrush clothing, faces, and bodies. They were making rock stars look cool, and charging rock-star prices.

When I walked into the studio and saw shirts, pants, and jackets with carefully-crafted, perfectly-executed psychedelic artwork for the likes of Bob Dylan, Led Zeppelin, Donovan, and Paul McCartney, I was over-the-moon with enthusiasm. These hippies had a work ethic and business acumen. The food was vegetarian, and the grass was not greener on the other side of the street.

Sweet memories of Oregon flooded my senses. It was the hippie vibe on commercial steroids; I thought if anyone could get Lloyd Cross and company to meet a deadline, it would be Star Rainbow.

Lloyd Cross had conquered Rainbow holograms, the process he was about to unveil at the convention in downtown Los Angeles. This would allow holographers to make holograms that could be illuminated by a 110-watt light bulb instead of a laser. Three-dimensional images could be reproduced on film and sealed between two sheets of plastic. Even a child could see the elusive 3-D object without much maneuvering. This was a huge scientific and commercial breakthrough.

There was a full color hologram of "Pristine," a San Francisco Cockette transvestite, and another one of Pam Brazier. There was a full color train and an art piece by Pim.

The San Francisco crew included an English girl named Lisa, along with Lilliana, who had taken part in our Sausalito infinity glasses revelry. Eddie Auswachs, the sharp producer I'd recently met poolside in Beverly Hills, had driven down to confer with Lloyd. In true hippie spirit, the genius inventor shared special effects shortcuts that could save Auswachs thousands of dollars. It was Old Home Week, a town reunion. Everyone was mellow, organic, and happy. After giving birth, Pam was skinny and confident, and after my trip to Hawaii I had put some of my Lloyd Cross-related New York angst behind me.

> Friday, January 11, 1974
>
> Had great pancake and stewed fruit breakfast. Made arrangements to see Silverbird concerning jewelry on Wednesday. Calling it Jupiter 5 really got her off. Challenge is my keyword. Got back to L.A. around 5 pm and dropped Pam & Lisa & Lilliana at the Hilton (downtown L.A.). Pim, Lloyd and I went to the office (my father's) and wrote information to be printed Saturday for the show. One for all—all for one.

On Saturday I called forty potential investors, inviting them to Lloyd's holography exhibit at the convention center in downtown L.A. This was before anyone used "art" and "downtown L.A." in the same sentence. Showing someone a hologram one-on-one might be exciting for a moment, but add a crowd of "ohhs" and "ahhs" and the checkbooks would fly open. At least, that was my hope.

As it turned out, the Westons, Lynn, and Dr. Ray, came to the show and Lynn said she was ready to invest. On Sunday, a Neiman Marcus executive brought his busty blond wife and their eight-year-old daughter. The little girl asked how much my necklace cost. Knowing that questions like that were impolite I replied, "I don't know. It was a gift." She then turned to a girl from the School of Holography. "How much did your necklace cost?" "Sixty dollars," she replied. "Oh," said the child with disgust, "is that all?"

For me, this was a typical Los Angeles reaction. Sure, investors might get excited and say they were going to invest, but rest assured, by the time they hit puberty they were already jaded and wary. They were going to do their due diligence before signing over penny one, and researching Lloyd Cross and the School of Holography could be problematic.

Star Rainbow had a gift for motivating people and keeping them on track. He was completely taken with integral or multiplex holograms, and with hologram jewelry and all the amazing applications we described. He was the only person I could think of who had the right personality to get Lloyd to deliver the goods. Maybe even on time.

Wednesday, January 16, 1974

It seems that we have a group of cosmic tribes and I am definitely part of San Francisco. Our ideals are high and our stubbornness immutable. In hippie terms, "tribes" is an accurate description for group connectivity. If you met someone and you shared the same vibe, you were kindred spirits—you belonged to the same tribe. And who doesn't want to belong?

Thursday, January 17, 1974

Drove to Laguna with a laid-back attitude. Star (Star Rainbow) loved The Lovely Lascivious Laser Lady (four original poems) and plans to publish them in Lemuria magazine. He intimidates me and I told him so. He reminds me of some of my college professors, and a little of Jerry Brandt. He's got a heavy vibe and has a tendency to scare.

Star Rainbow liked my hologram bracelets and decided to expand the Jupiter 5 collection. He told me to work with a Rainbow Kids' artist named Toni. We were supposed to come up with a fifteen-piece line. Toni would make special pieces and design, and I would launch the business. Luna, another artist, made me a special Laser Lady air-brushed t-shirt. I still have it and I still treasure it.

On the drive back to L.A., I realized how much my right eye was bothering me. Apparently, being mellow doesn't cure a virus in one's cornea. My nerves were shot and staying in Brentwood at my parents' house, even temporarily, wasn't helping. My eyes were screaming, "Get an apartment!"

Tuesday, January 29, 1974

Went to Dr. Wolston and had my right eye held open with a clothespin type gadget and curated; wiped virus off eye. Painful. Horrible. A black patch over the sick eye. Daddy called mother asking if I could come to his office and fill in because Ingrid wouldn't be there until 12. I felt obliged and blindly drove down there. Ingrid was there. Then, amazing an old favorite, familiar voice, Peter (Cookson). I drove blindly...saying, "Only another mile, you can make it." And I did. Peter looks great. Same elusive, no strings soul. He capped our visit with "Could you use some money?" I smiled and he wrote me a check for $200. Very nice and needed.

On February 1 I received a letter from Cecile Ruchin that did not mention my HCCA stock issue. I wrote her back immediately, explaining my position. I could easily align myself with McDonnell Douglas or TRW and form my own hologram marketing company.

Cecile's response was immediate. She assured me that she was seeing her lawyer on Thursday and I would definitely be issued stock—and we were definitely going to have an international holography/art exhibition. She wanted me to speak to well-known artist Bruce Nauman, and to Dr. Ralph Wuerker at TRW about creating holographic art pieces for the show.

There were too many things going on. A friend of Charlie Patton's named Leon Young suggested we form a company to mass-produce my holographic jewelry. Charlie had begun to realize that not everyone was his friend. He was known to liberally spread what should have been confidential information. And, if he was drunk, he had no filter. It was as if when he talked to the right person, that individual would tilt him upright and make his life worry-free and financially stable. If he was going to be a partner in Jupiter 5, we needed a company structure, and neither one of us had taken a course in basic business.

Tuesday, February 5, 1974

Leon Young called and Joan Nielson, who'd been making sketches of holographic jewelry, and I dashed to the Bank of America building at Wilshire and Beverly Drive. Jewish mafia, international hustlers, (Perel says) "He'd kill you for 27 cents" to Leon. He, Perel, Romanian type, inventor, and two other attorneys talked about building a plant to mass produce (hologram) jewelry in a "secret country" (Indonesia) "Don't show this to anyone else!"

This sounds like an episode of Rocky and Bullwinkle. Intrigue, foreigners talking in hushed, threatening tones to two young, naïve American women on a corner in Beverly Hills with blue skies and tall palm trees swaying in the breeze.

Needless to say, my friend Joan and I made a beeline for her house, leaving the creepy investors in the rear-view mirror.

Wednesday, February 13, 1974

Cecile (Ruchin) came in on Freelandia (airline). I was very depressed because Columbia didn't come through with an option (Revelation II). Cecile and I had a joyous reunion. Went to Alice's Restaurant in Malibu for drinks and catching up. Bill O'Hare raising $100,000 for new line of jewelry. On and on we fabricated our plans.

Thursday was Valentine's Day and my friend Alan Abrams, Joan Nielsen, and I headed to the Forum in Inglewood to see Bob Dylan and The Band. They opened with "Most Likely You Go Your Way and I'll Go Mine." Hearing songs like "Knockin' on Heaven's Door," "Blowin' in the Wind," "Just Like A Woman," and "The Times They Are A-Changin'" made it one of my all-time favorite concerts.

Friday morning it was back to business. I called Dr. Ralph Wuerker at TRW to confirm our afternoon meeting. I picked Cecile up and off we went to TRW, the giant aerospace company in Redondo Beach.

As a pioneer in laser physics and holography, Dr. Wuerker had secured a number of patents. TRW treasured him and the U.S. government needed him. He was a very attractive man in his forties with a family and community standing. Funding was not his issue; time was.

> Friday, February 15, 1974
>
> TRW—showed Dr. Wuerker rainbow holograms: cable car, sculpture Point Star, dragon ring. He was enthusiastic like a child, laughing and saying, "Heh? heh? heh?" Lovely man. Discussed Bicentennial, film trip to Rome (cost TRW $25,000 to send Wuerker and crew), (Simon) Ramo responsible and got money's worth in publicity. He let Cecile and I know that the government pays scientists very well. "So they have the proper incentive," he said. He's working on non-destructive testing—anything but art.

Ralph Wuerker was charming and very artistically inclined. Cecile and I spoke to him about producing educational, medical, and art holograms for HCCA. Even though I think we all enjoyed the meeting, I had the sense that Dr. Wuerker was much too well taken care of at TRW to get involved with projects that didn't have adequate funding. He appreciated what we were attempting to accomplish, but he wasn't interested in diverting his attention.

By the end of the week I was feeling like things were coming together. William, my astrologer, called to say that Lynn Lenau was recommending me for an AIP writing assignment in Dallas. My friend Patrick had just played a robot on the TV series *Kung Fu* and had been invited to the Gene Roddenberry Festival, which I felt was the right time and place to launch Jupiter 5.

> Sunday, February 17, 1974
>
> Fred (Butler) called from Cleveland. Read me the song he wrote for me. Very deep and committed to me. Says he's been faithful and have I been? He really believes we can make it and maybe we can.

To further complicate my life, I had been dating a wonderful black singer-songwriter named Fred Butler. He spent most of his time in Cleveland, but wanted to move to L.A. I'd met him at Warner Bros. Records and really liked him. In my diary I wrote, "He deserves the best, most devoted counterpart. When he walks in a room, the room lights up."

What, you may ask, was the problem? In a word: parents. My parents were conservative Republicans who cared deeply about what other people thought. My mother wouldn't even let me invite my high school class president over to study because he was Mexican. Growing up in a racist vacuum, then leaving it to experience the world, I realized pretty quickly that human beings needed to be judged individually, not painted with a single brush.

That said, I knew that if my parents found out I was not only dating Fred, but that we were in a serious relationship, I would lose the use of my father's office and the jobs he paid me to do. I would be disinherited, and there was no guarantee that Fred, talented and charismatic as he was, would be successful in the music business.

I sigh deeply when I think of Fred Butler. He was a normal person, and I was so emotionally damaged there was no way I could completely open my heart to anyone, with the possible exception of Richard Harris. For him I was still putty.

A young me in parent-pleasing mode, 1969.

My uncle, Arthur Lane with my father Richard Lane and his world-record Black Sea Bass caught off Catalina Island.

The Opening of the Paradise Ballroom with
The Eat Me Girl and me as a clown, 1972.

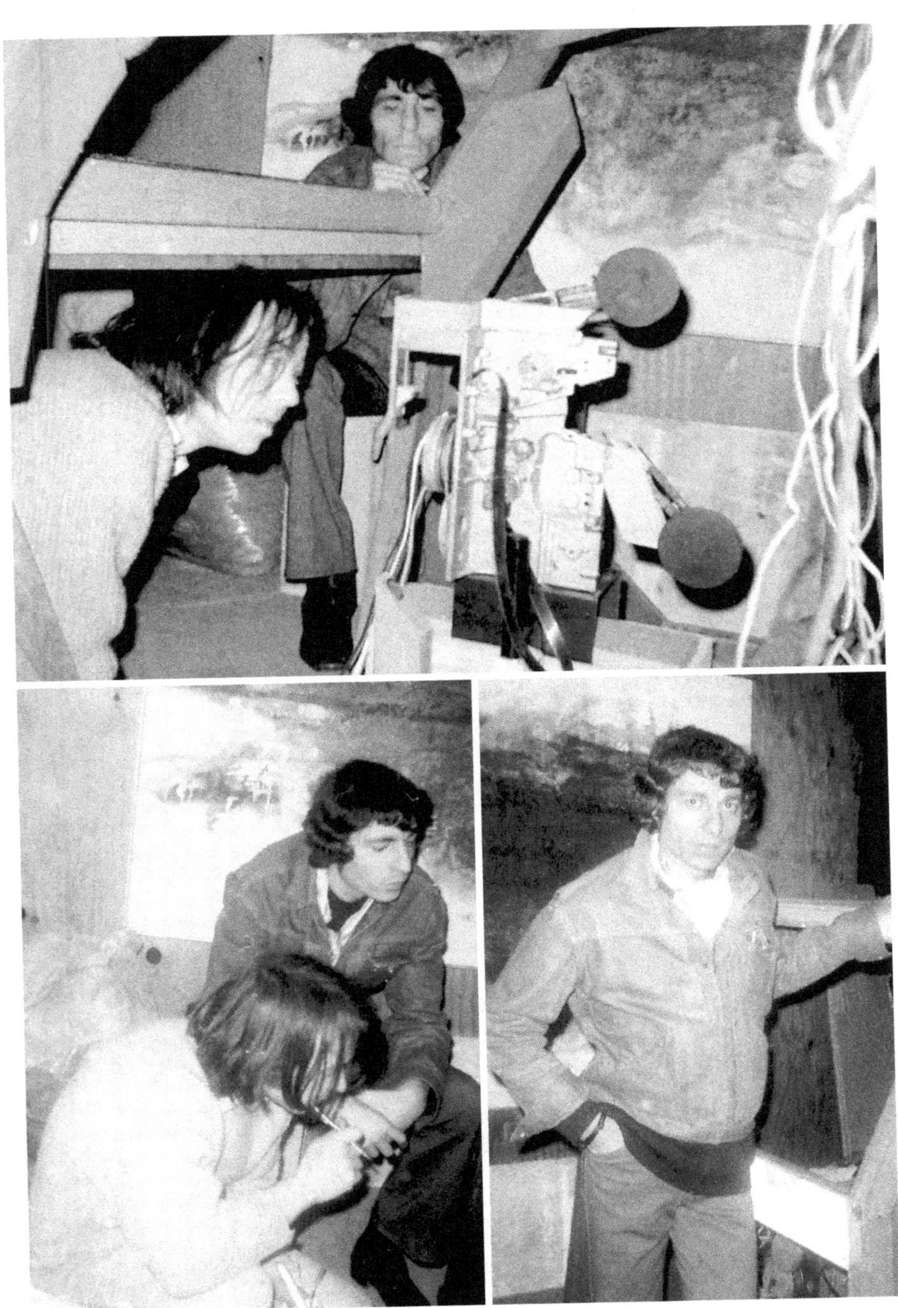

Lloyd Cross and Selwyn Lissack building the first multiplex camera and printer at the School of Holography in San Francisco, 1973.

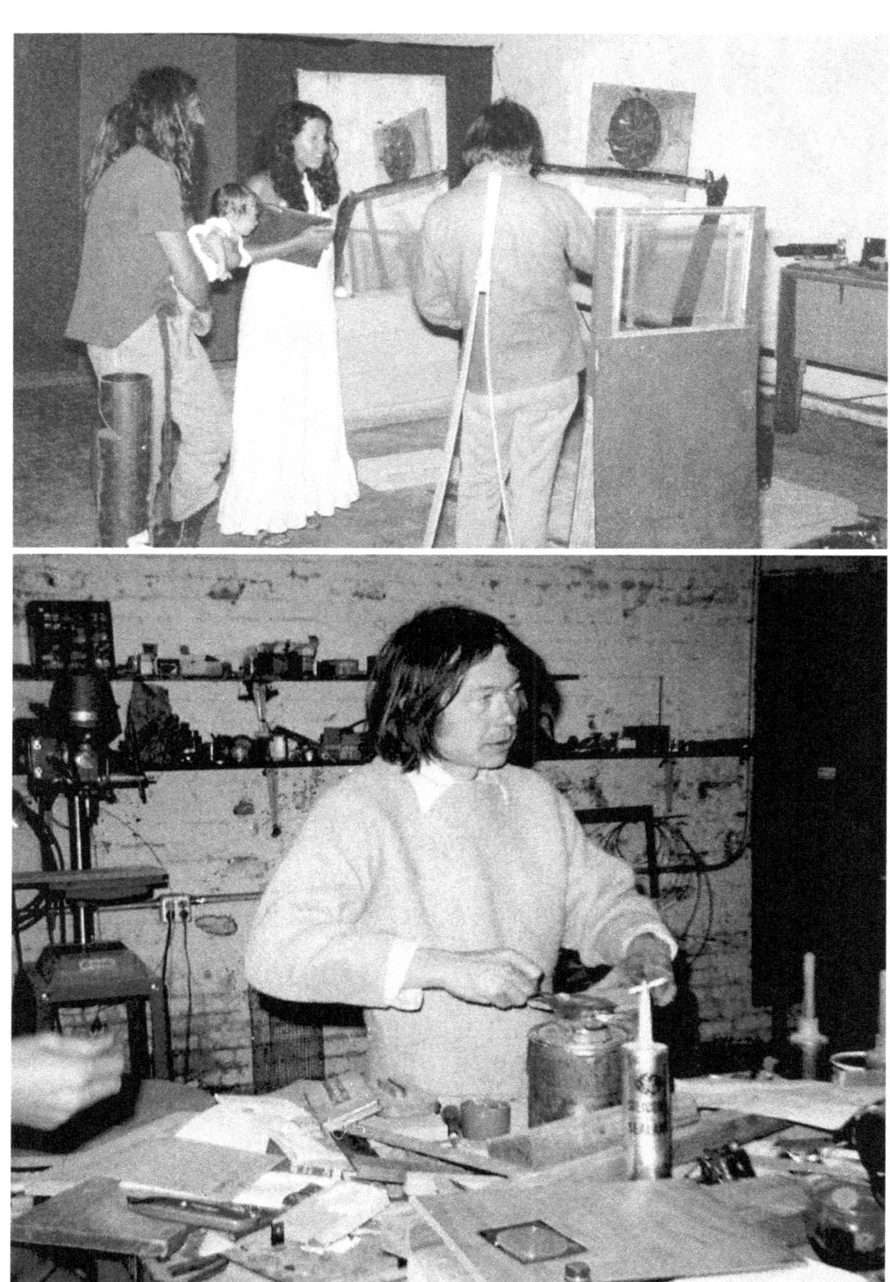

Yours truly, Lloyd Cross, and Barry Gott creating pioneering imagery at School of Holography in San Francisco.

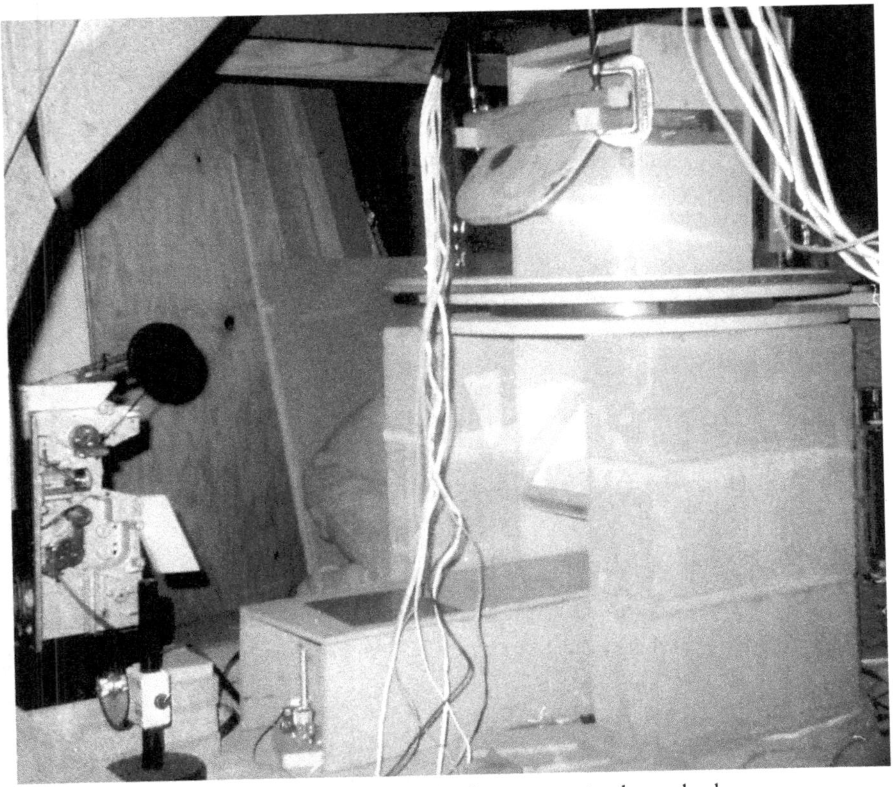

Creating a holographic multiplex camera in the early days required a lot of jerry-rigging.

The almost completed rotation stage and exposure film holder for the Dalí-Cooper 360° multiplex hologram.

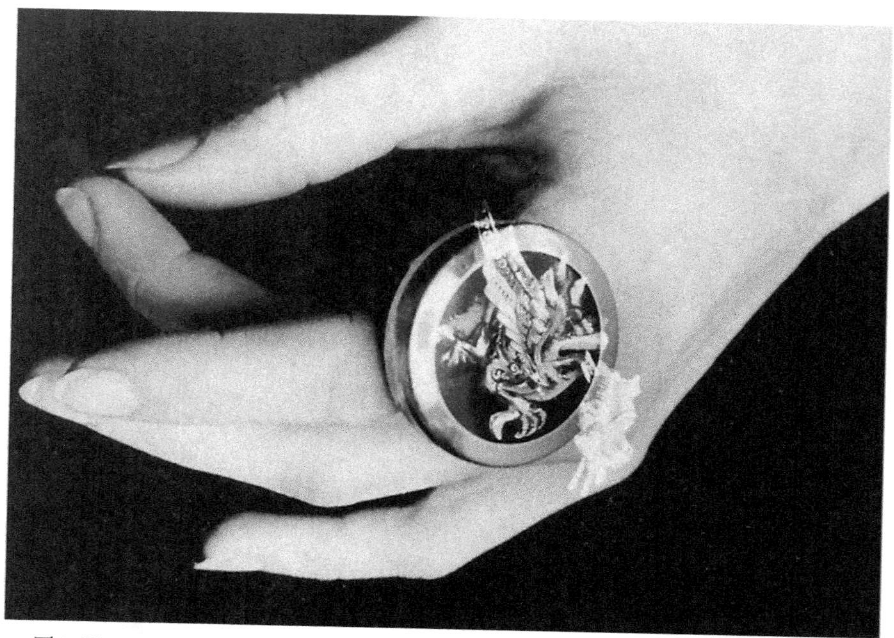

This Ysan Dragon ring was created by HCCA and shot by McDonnell Douglas.

Me and Cecile Ruchin modeling Jupiter 5 jewelry.

Percussionist Rocky Dzidzornu performs on conga drums using five different diffraction grating patterns and colors at our Welcome to Creation event, 1974.

Keystone Press / Alamy Stock Photo

Supermodel Donyale Luna loved Jupiter 5's Lumilar jewelry.

Me, Candy Clark, and screenwriter Ivan Moffat.

Mary Evans / STUDIOCANAL FILMS LTD / Alamy Stock Photo

Candy Clark on the set of *The Man Who Fell to Earth* with director Nic Roeg and David Bowie, 1976

CHAPTER 14

JUPITER 5

When Mi Flick arrived in Los Angeles from Seattle, she brought me a fabulous silk chiffon blouse that was perfect for many of my meetings—writing, selling jewelry, planning an exhibit for the 1976 Bicentennial, marketing multiplex holograms, or raising money for any and all of the above. My plate was full; however, I felt like the various threads were somehow connected. Like one of those games where one steel ball hits the next and the next and the next until they're all moving in unison.

> Tuesday, February 19, 1974
>
> Incredible day. Saw Peter MacGregor-Scott and successfully got him interested in using footage from his next film for a 360 degree advertising piece (Multiplex hologram). In the afternoon went by Andy Pfeffer's office (attorney at Mitchell, Silberberg & Knupp) with Cecile (HCCA). Had a fantastic session. Showed rainbow hologram of Pam with a puppy, a trolley car, and a hologram called Point Star. Very impressive meeting. Met with Darryl Presnel at Mann Theaters. He would like to get "mini movie" advertising into theaters.

Shelly Davis wanted to see me at MGM, so Cecile and I headed to Culver City. The more time I spent with holographers and projects relating to holography, the happier I was. I wasn't under contract to MGM or to Shelly's company, Now Productions. I was a member of the Writers Guild and an independent contractor. I knew Shelly wasn't happy about my distraction, but I wasn't happy about working without a consistent paycheck. Owning 30 percent of nothing is still nothing.

There is a simple reason that I kept working with and for Shelly: loyalty. He believed in me and in his own way he became a father figure, always at the ready with advice. He was an anchor. "You need to write! Go back to writing, not Jupiter 5!" Shelly was adamant, but then, so was I. When it came to holograms, people would whip out their wallets. There was no hesitation. Writing seemed to be one high and low after another, and there was always another hurdle, another group of people who had the power to green-light or crush you.

Meanwhile, 1976 was rapidly approaching, and if HCCA and Jupiter 5 were going to be a presence at the Bicentennial, we needed to write a proposal ASAP. Cecile, Charlie Patton, and I spent hours crafting a presentation that combined antiques, antique replicas, and holographic scenes. The exhibit would be a three-caravan traveling show, and the projected cost to transport one unit with a tractor was $4,000 per week. We needed both corporate as well as government sponsorship if we were going to raise the millions needed to mount this production. And on top of that, we needed holographers who could meet deadlines.

Jupiter 5 was quickly becoming the West Coast arm of HCCA. Just as I had been energized at Cecile's in New York, she was feeling the Promethean spirit of people on the West Coast. On Monday she spoke to Hugh Wyand at the Diffraction Grating Co. concerning Jupiter 5 and representing them.

Anyone who is unfamiliar with the term "holographic diffraction grating foil" need only to think of shiny kinetic gift wrap—those rolls of Christmas wrap that radiate kaleidoscopic spectral patterns when light hits them. Boxes for cereal and beer boxes featured printed holography to outshine other products on grocery shelves.

It was clear to Cecile, Charlie, and me that adhesive diffraction grating foil could be adhered to jewelry, cars, notebooks, clothing, musical instruments—anything you wanted to catch light and turn heads. It came in rolls, was pliable, and could be used to manufacture any number of products. Lumilar, as I called it, came with a silver or gold background. Later more colors were added. Using a laser, a variety of prismatic colors and geometric shapes were engineered. The material offered limitless possibilities and it was already in production. All you had to do was place an order.

On Friday Cecile and I went to the beach. There's nothing like salt water and sea air to bring out the information at the bottom of the well.

> Friday, February 22, 1974
>
> Cecile told me why HCCA stock hasn't been issued to me. "Selwyn thinks you're a dingbat and I've had to work, you don't know how hard I've had to work, to contain the situation. He wants to brain tap and then get rid of—eliminate—the person instead of including them." It relieved me to know that Selwyn was the sticky cog. Now I know it, and though he turns into a monster on occasion, I believe I see some of the dark feelings I felt around him. I don't think he likes people in general because he doesn't trust or like himself too much. Peter Burtness called from Santa Ynez—spoke to Cecile about his cassette system and holography.

My friend Joan Nielsen gave a dinner party for 135 friends. I made carrot cake. It was a wonderful party. I wore the designer blouse Mi gave me and met Michael Van Horn, the chief fashion illustrator at *Women's Wear Daily*. He flipped over my holographic jewelry. "You have to meet Beth Ann Krier at the *Times*! I'll call her." It was a night of networking, and my jewelry was the focal point.

I had managed to rent an apartment on the border of Brentwood and Santa Monica from an old woman with short grey hair that looked like she cut it with nail scissors who always seemed to appear in a house-dress. She wore orthopedic saddle shoes and her mouth turned down in a frozen frown. Plain and simple. To her, young people were ruining the world.

The apartment was one room with a bed and not much else. I remember it being tiny and backing up to an alley. I wasn't supposed to have guests, and of course, I did. It felt more like a college dorm than an apartment.

> Monday, February 25, 1974 10 AM
>
> Dr. Wuerker—TRW—Chris Outwater, Cecile, Alan (Abrams), and I had to sneak out the front door to avoid the landlady hanging clothes outside the backdoor. Chris Outwater picked us up and off we sped to TRW.

Chris Outwater was the antithesis of Lloyd Cross. He was a young, good looking, blond-haired, tennis-playing preppy. He had a passion for holography and approached the subject from an academic/business perspective. He was grounded and he expected to conduct his lab and work in a completely conventional, professional way. Unlike many of the people I was meeting, Chris wasn't trying to become an overnight millionaire. He was content to put one foot in front of the other and stay on solid ground.

That Monday at TRW, Cecile selected twenty-three of Dr. Wuerker's holograms for an exhibit at the Knoedler Gallery in New York, the eponymous showroom that represented Salvador Dalí and the Alice Cooper-Dalí holograms.

Once again, Dr. Ralph Wuerker was cordial and firm. He liked working at TRW, and he wasn't driven to become a superstar in the art world. He was well paid and secure. He could not have been nicer, but he wasn't interested in pie in the bicentennial sky. If we got funding to commission a hologram, great—otherwise, end of story.

Monday night Cecile and I met with Charlie Patton to discuss his participation in Jupiter 5. He was divorced with a young son and working full-time at a design studio to pay child support. There was no Kickstarter or GoFundMe. Seed money had to come from investors who were willing to take a leap of faith.

Somehow Charlie had become friends with Doris Duke, the tobacco heiress and renowned art collector. They spoke on the phone every week.

"Doris, this is going to change the world!" he'd implore. "X-rays will be 3-D! A doctor 'll be able to see what's going on inside your body. In real time!"

Then he'd put me on the phone and I'd use my powers of persuasion, but nothing moved the mountain. Doris listened and sounded fascinated, but she never invested.

That night Charlie had to agree that in spite of all his contacts, and there were many, no one was stepping up to invest. They loved his enthusiasm, but the financial rewards didn't outweigh the potential risks. Charlie reluctantly agreed, he needed to keep his day job until Jupiter 5 was capitalized.

Now that we'd sorted things out with him, Cecile and I had to find a gas station with less than a two-hour wait. The gas shortage made driving around L.A. a nightmare. Gas was being rationed, and although my father had a pump at his office, it was only for himself and his trucks. Running on fumes, Cecile and I finally threw in the towel and paid for a car wash so we could get seven gallons instead of the sanctioned five.

We spent every day meeting with people who would either order diffraction grating material from Jupiter 5, hire us to come up with a way they could use it to market their products, or commission a multiplex hologram.

One day we found ourselves in an unfamiliar area of Hollywood searching for a pair of avant-garde clothing designers named Chance and Loretta. They wanted to incorporate holographic diffraction gratings into their eclectic, space designs. They were talented Texans whose clothes retailed for between $100 and $300.

Finally, we came upon a little old house sandwiched between rundown apartment buildings on the outskirts of Hollywood. None of their neighbors—Chinese, Latino, Middle Eastern—spoke English.

"We've been praying for this material," Chance drawled. He told us he was asexual and he knew exactly where he was headed. Loretta was sweet, skinny, and the perfect model. They ordered yardage.

Wednesday, February 27, 1974

Waited and fretted because Zar (Laguna Beach Rainbow tribe) was late. He arrived with plans for first pyramid "high energy" structure. Silverbird into organic menu. Zar wants to use diffraction grating material on one wall 50′ x 60′, very positive. Cecile and I went to the office. Waiting for Joan told Charlie we're filing papers (Jupiter 5) and it would be better to eliminate necessity of his assuming responsibility for Jupiter 5 at this time. Cecile (HCCA) 50 percent, Linda 40 percent, and Joan if she functions as administrator/art director 10 percent.

Things were moving quickly, almost too quickly. I wrote Fred Butler a letter explaining that I was not willing to trade the innovative work I was doing for a relationship. He called, trying to pressure me, and I told him he was strong and he would have to bear with me. Saying "no" had always been difficult for me. I didn't want to lead Fred on, but I didn't want to be blunt and cause more drama. My parents fought on a daily basis, and I was shell shocked from so many years of tension. Trying to launch a company and be all things to all people was more pressure than I could take. My right eye was still bothering me.

Thursday, March 1, 1974

Met with Tim Small at Hughes [Aircraft]. Discussed Bicentennial, borrowing their equipment, and overall picture. Small, and as with many scientist types, likes artists. Met Chris Outwater at the Swiss Café and suggested he publish his book as "Hologram Press Publishing." Office. Late visit to Zox & Deb's in Venice. Zox = cosmic multimedia circus. Imaginative art, combination between 50s rock and 70s space. Dea came home—said Beth Ann Krier wants to write a feature story on me for L.A. Times—all inspired by Michael Van Horn at Joan's party.

Friday morning I called Beth Ann Krier at the *L.A. Times* and made an appointment to meet with her the following week. Then Cecile and I met with Barry Taff from the Parapsychology Laboratory at UCLA. He had a doctorate in psychophysiology with a minor in biomedical engineering. In my diary I wrote:

> Barry Taff, brilliant psychic technologist. Works with Thelma Moss, Uri Geller. Small, talks fast, been ripped off by Hollywood ESP shows. Inventor. Needs funding. Government hears—use computer or can photograph thoughts holographically—prove our thoughts have form—mass. Government freaked—don't want the responsibility.

Cecile, who looked like an alien wearing dark wraparound sunglasses with a jumpsuit and turban day and night, had a penchant for people who were able to plug into the ether. I, of course, found these things fascinating, ever the eager student.

Cecile and I met with my old friend David Lewis, who had been a pioneer in television. He looked at our prospectus and found it too broad. Hindsight being 20/20, had we simply asked for funding for holographic jewelry, he thought we would have had investors. Even my close friend Lynn Weston seemed reluctant to risk the capital she'd offered at the convention center. The irony was that everywhere I went people stopped to ask me about the jewelry. Inevitably they wanted to buy it, or in many cases, sell it for Jupiter 5.

That Sunday I met with Joan Nielsen to iron out her interest in the company. Joan was one of the most talented people I'd ever known. Talented as an artist, an illustrator, a caterer, and as a writer. She was a willowy brunette with style for miles. Grounded and organized. We'd been friends since I'd helped her get a job at the crazy Paradise Ballroom. Now we were about to launch a company that promised to revolutionize marketing and add new dimensions to fashion. Yes, she was in for 10 percent.

On Monday, March 4, 1974, we filed the Jupiter 5 fictitious business certificate and paid to run the name in the *Beverly Hills Independent* newspaper for four weeks. I spent thirty-three dollars. If only I had known the real cost.

CHAPTER 15

BIRDS OF A FEATHER

Jupiter 5 was now a legal entity, and Cecile and I had to streamline our endeavors. We were dealing with a cast of players who had very specific, self-interested agendas. I was trusting and idealistic, always seeing the glass half full. This allowed me to view opportunities through rose-colored glasses and to avoid unpleasant realities such as: What is your business plan?

Tallie Cochrane, the Southern belle porno producer and her actor husband, Patrick Wright, whom I'd met working on Shelly Davis' TV treatment, wanted Mr. and Mrs. Flash Gordon costumes. Wright had played a robot on the eponymous series and he was obsessed. The couple needed a futuristic costume designer and on the face of it, Chance and Loretta were the right people for the job. What I hadn't realized was that Chance considered himself a star, and unless he was paid a major studio designer's rate, he wasn't interested.

Feeling like I should seek out an alternative, I was introduced to Bjo Trimble, a science fiction fan who was known to make superhero costumes. The Los Angeles home she shared with her husband John was large, with rooms that were usually reserved for living or dining being used as workshop spaces. My first impression of Bjo cast her as a high-strung ball of energy, a multitasker on a cosmic mission. Her looks may have been plain, but her imagination and love of science fiction put her in the room with Gene Roddenberry and Forrest J. Ackerman. Bjo was receptive, even excited about making the Flash Gordon costumes.

The best news of all was that Jupiter 5 was now officially representing the Diffraction Grating Co. We were marketing their product while creating products of our own. We sent letters to potential investors, as well as to people who could purchase quantities of material, such as George Lucas.

> Wednesday, March 6, 1974
>
> Silverbird consulted her lights and decided not to make holographic jewelry, but to concentrate on pyramid, food, etc. Met Ron Goldstein, Eddie Auswach's friend. Congenial, spiritually oriented household—Shangri-La Ranch. Two horses, one pony. Ocean block away. Mountains. Lloyd (Cross), Pam (Brazier), Gary (Adams), Dave (Schmidt), etc. at the pyramid house.

Shangri-La Ranch in Malibu has become an iconic setting for musicians, most notably in 1975 and 1976, when The Band and Bob Dylan lived and recorded there. Today, Def Jam Recordings co-founder and music producer Rick Rubin owns the property and records everyone from Jay-Z to Kanye West to Adele in the two original studios.

I spent a good deal of quality time at Shangri-La Ranch because Eddie Auswach's friend Ron Goldstein either owned or was leasing the property and it was the perfect place in 1974 for the San Francisco-Lloyd-Cross-School-of-Holography tribe to come together with the Laguna-Beach-Rainbow tribe. The Laguna group was interested in building a pyramid, while the holography people were contemplating holographic windows, works of art and products that could be mass marketed.

While Cecile and I were basking in the Malibu sunshine and trying to buy enough gas to get from point A to point B and back again, Selwyn was lighting up the art world in NYC with commissions from both Andy Warhol and David Bowie.

Thursday, March 7, 1974

New York doing Integral hologram of Candy Darling, Warhol's star who's dying of cancer. Contracted a viral infection in the hospital. Selwyn says, if necessary they'll send Billy Maynard to the hospital to film her in bed. End result—Candy jumps out of a giant candy box. HCCA just did a job for David Bowie. Heavy action in New York. Money wants to invest!

While the East Coast was filming Candy Darling in 16 mm to be turned into a 360-degree moving hologram, Cecile and I were on our way to shoot footage of a Freelandia DC-8 aircraft at the Mercury Landing Building in Culver City, California.

The startlingly yellow DC-8 with the open hand insignia on its tail was surrounded by over a hundred partying travelers celebrating with champagne and cake. Cecile had set this up and she was determined to direct the cameraman to shoot completely around the plane. To accomplish this, the DC-8 began rotating, jet engines blasting, literally blowing the cameraman off the tarmac.

The brilliance of Freelandia was that it was a club. For $25 you joined and then you got to enjoy a fabulous party wherever they flew—Newark to L.A. or Newark to Chicago or Mazatlan or Brussels. Due to the fuel shortage, major airline prices were sky high, but not Freelandia's. Flights cost 50 percent less and were 100 percent more fun. Even the stewardesses and stewards wore uniforms with Flash Gordon flair.

Billed as a nonprofit, revenue was donated to a number of charities. Passengers grazed on organic food and enjoyed all out partying while airborne. Kenny Moss, a young self-made Wall Street millionaire, had created the ultimate winged Mardi Gras. It took off in August 1973 and landed forevermore in April 1974.

The footage had been intended as a marketing device for Freelandia. That possibility evaporated when the company closed its doors. The cameraman, however, survived the wind tunnel and gave us his card for future assignments. He liked the idea of a three-dimensional object floating in a cylinder.

On Friday, March 8, Cecile and I were invited downtown to meet *L.A. Times* staff writer Beth Ann Krier. "Hello," she greeted us, taking in our colorful kinetic jewelry. Beth Ann exuded warmth with the wide-eyed excitement of someone seeing a Maui sunset for the first time. Slightly plump and brimming with unbridled enthusiasm, she christened us the cosmic duo and called designer Rudi Gernreich.

"Rudi, I have these two interesting girls in my office and I think you should meet them." She attempted to describe our luminescent, light-sensitive holographic jewelry. It was an exhilarating meeting ending with Beth Ann expressing her interest in writing an article about it. Giving us an introduction to Rudi Gernreich, a futurist himself and one of the most daring and iconic American fashion designers of the 20th century, gave Cecile and me the feeling that we were airborne and nearing the landing strip. We had almost arrived.

> Friday, March 8, 1974
>
> Sitting at American Airlines waiting for Selwyn, and Jody (Burns) to walk out. A million Selwyn thoughts, but first, shock—they're renting a car. Money = freedom. Selwyn had an interesting beard. Very cool. Holds far back. Shy. Jody warm, sense of humor, wise.

New York had flown out to L.A., and it was important to me to orchestrate smooth sailing. Our first stop was The Studio Grill, a film industry haunt owned by my good friend, Ardison Phillips.

It was intimate, with a handful of banquettes, a few tables and chairs, a bar, and an extensive wine list. When we arrived Ardison was drunk and not communicating well. I'd given him a big build up, telling the holographers that he was responsible for a brilliant laser infinity installation at Cal Tech. I was embarrassed because Selwyn and Jody viewed Angelinos as lightweights, self-indulgent, and narcissistic.

It seemed that instead of filling an HCCA multiplex order, Lloyd Cross used some of Selwyn's holographic film to make rainbow holograms. Nothing seemed to be going smoothly.

Later, Jody Burns came by my tiny apartment and we discussed potential HCCA people and projects. I could tell that he was very skeptical. He wanted facts. How was I going to raise money? What was the Jupiter 5 business plan? Exactly what had Cecile and I accomplished since her arrival? Jody was young and attractive in an East Coast, Ivy League way. He had obviously paid attention in college, understanding both the science of holography and the mechanics of business.

Saturday morning Selwyn, Cecile, Jody, and I met Chris Outwater for breakfast at Nate 'N Al's in Beverly Hills. It was our intention to discuss Chris setting up his own Multiplex facility. Cecile told her partners, "He's a sleeper," meaning that when you met him he wasn't going to set the world on fire, but yes, he could do the work and he was dependable.

> Saturday, March 9, 1974
>
> Lloyd (Cross) and Al leaving for Venice to find a (work) space. Drive by Charlie's for good measure. I leave a note and just as we're driving off, Lloyd's van drives up and Charlie (Patton) gets out. They were obviously discussing business.

As I mentioned, everyone seemed to have his own agenda. Lloyd felt that he had invented Integral or Multiplex holography and he wanted to patent it and maintain control. HCCA was OK with that in principle, except Lloyd was known to go off on tangents and miss deadlines. In their view, the San Francisco tribe was more hippie-minded than professional. Then there was Charlie Patton, who was obsessed with pioneering holography as an art form. He too was jockeying for a position.

We were all heading out to Shangri-La Ranch in Malibu, so I quickly made another carrot cake. Apparently, I was known for my carrot cake. I'll admit, it was a crowd pleaser, and great fuel for cosmic thought, but unfortunately, on this day nothing sweet and delicious could cut the tension.

A business meeting between Lloyd Cross, Dave Schmidt, Gary Adams—the San Francisco contingent—and Selwyn Lissack, Cecile

Ruchin and Jody Burns—the New York business professionals— turned into a mental prizefight. And of course, in the background, in keeping with the Malibu setting, there were several beautiful nymphets sharing juicy rock star tales while sunbathing.

As the parties were engaging in serious discussions about building a business around multiplex or integral holography, Pam Brazier and Charlie Patton emerged from the sauna. No need for cover-ups or towels here. They stood watching the serious debaters and finally Cecile snapped.

"Come on!" she barked. "We've got to get to the point!"

Lloyd Cross and his San Francisco cohorts, bright and committed as they were to forging a lucrative business, were equally committed to enjoying every minute at Shangri-La Ranch—lounging in the sauna, relaxing on a hillside overlooking the idyllic blue Pacific, and living in the moment.

It became clear that these two factions were oil and water. They both wanted the same outcome—to be in charge—to live the good life—but their approaches were diametrically opposed. New York didn't trust San Francisco and San Francisco didn't trust New York.

I drove back to Santa Monica thinking about the disparate factions vying for position and the dynamic documentary this would make. I had been having discussions with Gertrude Ross Marks, a Golden Globe winning writer and documentary film producer, about this very subject. Gertrude and I had gotten to know one another at Writers Guild meetings and screenings. She admired my hologram jewelry and wanted to buy it. I told her about the documentary proposal I'd submitted to AFI, and we discussed the possibility of expanding it. HCCA would co-produce alongside her production company, Mentor Productions.

Sunday morning my phone rang. It was Fred Butler calling from Cleveland, his voice strained and filled with dread. Harry Womack had been murdered. Pat, Harry Womack's girlfriend, had stabbed him to death.

"What happened? "

"Harry wanted to boogie with Bobby..."

Fred was crying. He was close to the whole Womack family. Originally from Cleveland, they had come out of gospel before going into soul and R&B. They were close-knit, with Bobby Womack's guitar skills, songwriting success, and deep, romantic voice making him the undisputed star. Sadly, the night his brother Harry was murdered, Bobby was performing on stage in Seattle.

I'd spent time with them. I'd partied with them. I knew Pat and Harry. There was a beautiful child. I didn't know what to say. "Can you pick me up at the airport Monday?"

"Of course."

I can't understand it. Happened during a full moon. They'd just taken Bobby Womack's billboard off Sunset strip. What an inexplicable tragedy. It knocked the wind out of me. I had never known anyone who'd been murdered or killed someone before. Fred was in shock. And so was I.

The following smoggy Monday afternoon, I picked Fred Butler up at LAX and drove him to Friendly Womack's house near downtown L.A. Inside, voices were vacant, hushed. It felt like stepping into an Edvard Munch painting, so I was glad I had to pay my respects and excuse myself to meet Gertrude Marks and Ed Penney.

Gertrude was older, wiser, married with grown children—what you would call "well put together." She and her Mentor Productions partner, Ed Penney, had been nominated for an Oscar for their documentary feature *Walls of Fire*.

I joined them along with Cecile at Café Figuero in West Hollywood to discuss a joint venture between HCCA and Mentor Productions. Names of holographers in England, France, and Russia were tossed out. Why limit ourselves to the U.S.?

After this exhilarating meeting, I was being promised stock in HCCA once again. Cecile had explained the stock delay by blaming Selwyn. In retrospect, several things are clear. I was very naïve. I took what people told me at face value. And I would do almost anything to avoid a direct confrontation.

Cecile was going back to New York in a couple of days so we needed to make haste. Renowned fashion designer Rudi Gernreich had responded to Beth Ann Krier's introduction and he wanted to meet the cosmic duo.

Tuesday, March 12, 1974

Rudi Gernreich—8460 Santa Monica Blvd.

Gernreich is a short, well-constructed man with large brown eyes that reflect childlike enthusiasm. He loved Lumilar and is turning us on to Ed Ruscha and Larry Bell. Couldn't have been better!

As an avant-garde designer, Rudi Gernreich was first to use cutouts and plastic, and make comfortable wireless swimsuits, even the topless monokini—was the perfect person to use kaleidoscopic Lumilar foil. Rudi was fascinated with Jupiter 5 jewelry and wanted to buy yards of several Lumilar patterns.

Selwyn was now ready to sell rolls of Lumilar in New York, and since I was the one who'd brought this to everyone's attention, I felt that I should be given credit for connecting key people with Jupiter 5 and in turn HCCA. We were moving at warp speed, using the top floor of my father's office as well as my studio apartment as a base and ignoring Mrs. Mio, my nasty landlady.

I had just snapped a photo of Cecile with Chris Outwater when Mrs. Mio walked up, handed me an eviction notice, and walked off with her little dog following at her heels. My friend Alan had predicted this would happen and he was right. Cecile was not particularly upset because she was leaving in two days, but I was devastated.

Wednesday, March 13, 1974

Really down. Can't write to Rudi (Gernreich). Go back (to apartment) to find that Cecile has been invited to Laguna—a special and well-earned invitation. I felt lousy so went to Joan's (Nielsen) to see Jupiter 5 logo. Once again she wasn't satisfied. We both lamented our current living conditions and projected into the bright future. In many ways I felt badly that I hadn't been invited to Laguna, but I rationalized this as my need for growth.

Thursday morning Cecile and I were supposed to have a meeting with Gertrude Marks and Ed Penney, but she was still in Laguna. No problem. I went, giving the producers a copy of my outline. Ed Penney had put something together to prove that he was onboard. Their proposed budget was $300,000—much higher than I had anticipated—however it included trips and crews to capture the global pioneers of holography. The meeting went well.

When I arrived back at the apartment, I found Cecile and Silverbird. We were late for her flight so we tossed her things in my car and raced to LAX. Cecile rushed inside and rushed back out. "Jody said he bought me a ticket, but he didn't."

New York and L.A. were definitely not in sync. Cecile spent the night, which gave us time to discuss business. Friday I dropped her off at LAX and drove to Burbank Studios to see *Walls of Fire*, Mentor Productions' documentary.

The film examined the history of Mexican murals, exploring twenty years of David Alfaro Siqueiros, Diego Rivera, and Jose Clemente Orazco. It was a brilliant and revealing ninety minutes. To make a documentary on holography with documentarians of this caliber would be a gift from the heavens.

I went by the office and my father's secretary told me that a man from Dun and Bradstreet had just left, and she gave me a great recommendation. In the 1970s a good rating from Dun & Bradstreet was everything. Investors trusted them, and now I was being given their stamp of approval.

Good news, just like bad news, is contagious. Peter Cookson called and asked me to meet him at Café Figuero. He said he'd invest $700 in Jupiter 5, and to get started he handed me a check for $200. When he married actress Beatrice Straight, he became a producer and businessman. To me he was a mentor and an entrepreneur willing to gamble on *Revelation II* and now Jupiter 5. Peter was well aware of the fact that you couldn't launch a business on ideas and air. If I was going to succeed I needed to buy diffraction grating material and Plexiglass for bracelets. To succeed I needed products to sell.

Sunday, March 17, 1974

Sent Lumilar price list to Tom Lynott at Reinell. Jody Burns had a long talk with Chris (Outwater) who's making an 18" x 24" hologram for JBL—to be put on new speaker compressor. Lloyd's (Cross) face turned pale when Chris told him he'd been talking to "the people from New York". Chris will teach (holography) for $30 an hour at the School of Holography in Santa Monica that Lloyd is setting up. As Chris says, "The price of a good tennis lesson."

I needed to sort things out with Fred Butler so I agreed to accompany him to Sunset Sound, the studio where Bobby Womack was recording his new album. After his brother Harry's funeral, the singer's aging Cleveland relatives had stayed over and come to the session. So far they'd listened to Boogie Jones forty times. Bobby himself was MIA.

Finally, Bobby appeared…strutting…staggering…weaving through angel-dusted emotions over to Marty, the two-by-four balding Jewish producer, waving his arms and braying, "When my woman calls, you let me talk to her! Don't you ever hang up when it's my woman. You understand me?! I don't care what I'm doing."

All the aging black ladies gawked. They were enjoying the show. The tunes played on and Boogie Jones played for the fiftieth time. My right eye hurt and had a burning sensation. I wanted to leave and Scorpio Fred wanted to stay. We were locking horns until the musicians stopped and discovered my holographic jewelry. I got mobbed and handed out Jupiter 5 business cards. It was 3:30 am and my energy had returned. I was happy. I awed them, including Mr. Bobby Womack himself.

On Monday I saw the eye doctor. He told me my infection was back and if I didn't take proper care of it I'd be blind in my right eye.

Romantic pressure. Business pressure. Pressure. I needed to decompress or go blind. I'd been evicted and I didn't know where I was going to move. In the midst of this I discovered that Chance of designers

Chance and Loretta had written directly to the Diffraction Co. to buy Lumilar. Hugh Wyand caught it because I had alerted him to Chance as a potential client.

This was yet another blow to my emotional make-up. I'd given the designers material to experiment with, and instead of placing an order with me they thought they could save a few dollars by going behind my back.

Fred called to tell me that Bobby Womack was cutting two of his songs, and his friend Reynaldo had sold two hologram rings after his night club act. The circus wouldn't stop. I spoke to Cecile and wrote the following in my diary:

> Dalí won't fly. He sails from New York to the continent and reads Scientific American magazines galore. He spends his time growing which is why he won't waste time with anyone. He is a contemporary Renaissance man

CHAPTER 16

THE BEST-LAID PLANS

Monday, March 25, 1974

Unexpected trip to Coherent Radiation and Spectra Physics. Got up at 6 am and rushed to get ready. Barely made it to the plane—got to rainy San Francisco after nervous Charlie (Patton) had a few beers and chain-smoked Kools. Met Ian Knight, Led Zeppelin's special effects man, and Allen Owen from Showco. I was the driver and note taker. White-coated physicist at Spectra Physics humbly asked, "Who would like this diaelectric mirror?" "I would!" I said without hesitating. The men were shocked when I was handed a round 36" mirrorized gold piece of very valuable technical equipment. They were just too slow. This is not a dream! Keep saying that—at 3 AM eating a S.F. French bread cheddar cheese sandwich and pulsating from a miracle: ACADEMY AWARDS TICKETS with good seats.

Getting Academy Awards tickets has always been difficult because they're non-transferable. And, if the Academy finds out you gave them away, you risk losing your membership. I got them from a friend who had a family emergency. He'd already paid for them and he didn't want them to go to waste.

Bjo Trimble volunteered to make my dress—a gown, if you will. All I had to do was provide the fabric. I wanted Jupiter 5 holographic jewelry and strategically placed Lumilar cutouts to set a new fashion trend—to wow people! I spoke to Rudi Gernreich, who told me he was going to the Oscars. Everything was lining up nicely. Everything except the virus in my cornea. I had an appointment with a specialist at the Jules Stein Eye Clinic who told me not to sit in direct sun for one year. Apparently, the virus liked heat, and even though it might die down, it would never disappear completely.

> Friday, March 29, 1974
>
> Fitting at Bjo's. Met Roto Rooter Cosmic Band. All loved Lumilar. Went to office. Worked. Evading Fred (Butler) who's been left at Bobby Womack's large Hollywood Hills home with wolf and a freezer full of frozen meat. Fred's freaked. He wants me to come rescue him. I won't! I go to Filmex with Alan (Abrams) to see Orson Welles "Fake" and an English film with Ringo who was supposed to show, but didn't. Took $500 to Meta Noia for our share of jewelry manufacturing. Nick (Franzosa) couldn't wait to tell me: Meta Noia is going to sign a contract with Paramount tying up exclusive rights to all holographic representations of Star Trek subject matter. I was shocked since Cecile and I had been discussing it with him. I was going to contact the Paramount legal department, but didn't because Nick said he'd take care of it. Instead, he saw an opportunity to do exactly what Cecile and I had already discussed with Bjo Trimble and Chris Outwater. Paramount man won't be back til Monday.

Once again, everyone was waiting for a crack in the fabric of opportunity. What seemed to be happening was if you shared an idea it was no longer yours. It had been offered to the universal ear, and now there was a contract on the table and someone else's name had been substituted for ours.

It's easy to see a pattern emerging. Everything to do with holography and making money had to do with two things: who got there first, and who got his or her name on the dotted line first. Cecile and I had approached Paramount regarding licensing holographic products for *Star Trek*. Now, the person we hired to manufacture holographic jewelry was trying to get the license for himself. Never a dull moment.

On Saturday I went to another screening at Filmex and ran into the arrogant designing team of Chance and Loretta. A heated confrontation ensued, with Chance hysterically insisting on dealing directly with the Diffraction Co. Finally I told him Jupiter 5 was not raising the price to include our commission so I would ask the company to sell directly to him and pay us separately. At least one problem had been resolved.

Sunday I spent three hours on the roof of Meta Noia a design and marketing company for events that Nick Franzosa had formed. Nick and I were joined by Helen Nielsen, Joan's mother, who happened to be a publicist. Nick was insisting that Meta Noia, should hold the license to holographic *Star Trek* products with Paramount. Helen Nielsen had originally introduced me to Nick, so she became the voice of reason. Plus her daughter Joan owned 10 percent of Jupiter 5.

"Nick," Helen insisted, "we think you need to speak to Cecile in New York. Let's clear this up once and for all."

Helen and I listened in on another extension as Cecile told Nick the same things we had. He changed his tune when he realized that we were the ones supplying the holograms and holographic material, and if he didn't have us onboard he would have a very difficult time filling orders. He grudgingly agreed to make it a joint venture. Once we got that sorted out, I spoke to Cecile, who told me that the Diffraction Co. had hired Dun & Bradstreet to check our credit. Naturally, the report on Jupiter 5 was short. The report on HCCA listed them as manufacturers. We passed muster. We had been validated. We were approved for credit.

Monday I raced around tying up loose ends. Shelly Davis had asked me to write a treatment for a movie of the week dealing with a compulsive gambler. I was attempting to do that in between having a meeting with Nick Franzosa at Meta Noia, and an Oscar gown fitting with Bjo Trimble.

In 1974 the Academy Awards were held at the Dorothy Chandler Pavilion in downtown Los Angeles. They were much grander than when my friend Bruce Bauer and I sneaked into the ceremony at the Santa Monica Civic auditorium when we were juniors in high school. Wendell Corey was president of the Academy, and since we were close friends of his children, Bruce and I thought that if we got caught he would come to our rescue. Bruce had a tuxedo and I borrowed a silk chiffon gown and a fox fur trimmed coat from my mother, who believed we had tickets.

Sneaking in only heightened the experience. Seeing young Sophia Loren leaning against a door conversing with Carlo Ponti, her diminutive husband, took our breath away. The thrill of this youthful night at the Oscars triggered my lifelong commitment to the arts, especially film. For me, getting Oscar tickets was an affirmation. I'd been there twice before, once with tickets and once without. Now I was being given an opportunity to take fashion to the next level simply by mingling with the crowd. Oscar morning was spent driving my white Oldsmobile Toronado, a wonderful hand-me-down from my father, to Malibu to have Lumilar lightning bolts adhered to the front doors. The futuristic car had revolutionary front-wheel drive and metal headlight covers that opened and closed like eyelids.

I'll never forget that day, driving to the Pyramid house to pick up my lightning bolts…which weren't ready. Silverbird explained how they'd stayed up all night mathematically plotting the most powerful bolt. They'd managed to cut a pattern for one. Silver hurried, finished it, and put it on the car. That was the first distressing note, and naturally their excuse was God.

When I arrived at Bjo Trimbles, I found an innocuous writer judging space-age poetry and making erudite evaluations while Bjo attempted to chase her children, a cement truck, and finish my dress, which was coming together like a stage costume.

By the time I saw the shape it was taking it was too late. My date, Alan Abrams, arrived looking handsome as any young movie star.

The fabric was lavender and, in keeping with spring, it had a satin bodice and skirt with a layer of synthetic lavender chiffon over it. The outer layer was meant to have Lumilar cutouts sewn onto it. But they weren't. They hadn't been sewn. Hearts and stars and circles had been peeled off adhesive backing and stuck on all over the dress. Only they didn't stick—the adhesive was incompatible with the slick fabric and they fell off. And the only thing I could do was keep pressing on more hearts and stars.

I paced, glued fake nails on, peeled them off and re-glued them. I was going crazy. Academy Awards tickets were almost impossible to get, and here I'd managed to put so much in place, and my dress wasn't finished.

Finally Bjo brought it out. What?! It looked like a glorified prom dress from the 1950s. I put it on and died a thousand deaths. It was two sizes too big. I looked like an aging prom queen with poofy sleeves and no style, and there was no time for a do-over. Alan looked handsome and cool in his tux and I looked like a frumpy lavender confection. I made him stop at a drugstore so I could rip off the fake nails and cover the uneven mountains of glue with nail polish and glitter. We made it just in time and began our walk down the red carpet.

The bleachers were filled with fans screaming for a real celebrity when I came upon my old friend, broadcaster and gossip columnist Rona Barrett.

"Why Linda," Rona said, mic in hand and with friendly amazement. "I almost didn't recognize you, you've lost so much weight."

"It's the dress!" I pleaded. "I had it made. It's too big..."

Needless to say, my plan was completely foiled. I was in a foul mood. Our seats weren't arctic, but they weren't close to the stage either. We were surrounded by craftspeople and their spouses. The real action was up front, and since we'd arrived so late none of those people had even noticed my Jupiter 5 jewelry.

I wanted to close my eyes and disappear, but that wasn't possible. Glenda Jackson won Best Actress for *A Touch of Class*, and Jack Lemon picked up a Best Actor statue for *Save the Tiger*. Gertrude Marks and Ed Penney lost Best Documentary Feature to *The Great American Cowboy*. Finally, *The Sting* won for Best Picture and the ceremony ended.

As the crowd poured up the aisle towards the exit, my die cuts came off on them. I was miserable and Alan was mad at me for being so annoyed and probably, annoying. We went to Tommy's, a famous burger joint a few blocks away, and called it a night.

> Wednesday, April 3, 1974
>
> Rudi Gernreich—a lovely man of the first and highest order. Gave him diffraction (Lumilar) samples, Kirlian material and holography material. He's going to Seattle to promo his new perfume so I gave him Mi Flick's numbers.

Andy Pfeffer's office (Mitchell, Silberberg & Knupp). Nick (Franzosa) and his tacky attorneys in their Robert Hall suits came to the plush MSK offices and took notes, said nothing. Andy, Nick, and I came up with a viable deal. HCCA-Jupiter 5 receives all manufacturing of Star Trek holograms and 10 percent of net. And a larger percentage if we put up money. Andy sending it to Nick's attorneys in the morning. We agree and go to Paramount.

That Friday I signed the agreement with Nick Franzosa for the Meta Noia-Jupiter 5-Paramount licensing deal to manufacture holographic *Star Trek* products. At 2 pm Nick and I met with Lou Middling at Paramount. He was a wonderful gnome-like man in his sixties who seemed to be rooting for us. We took him outside to see the Ysan dragon hologram ring and he became animated. The more excited the Paramount executive became, the more enthusiastic Nick and I became.

"You have to know the blind spots too—remember that," he warned us. " I want you to succeed."

"Absolutely!" we agreed. Nick and I were ecstatic. We'd made it over the first hurdle, and Mr. Middling obviously liked us. Now we had to raise the licensing money. It would cost $2,500 to license one facet of *Star Trek* for holographic replication and $25,000 to license all of it. We also agreed to pay Paramount a 5 percent royalty, but that would come later.

My focus was on Equicon, which just happened to be co-chaired by John Trimble, Bjo's husband. Equicon was a convention for people who were interested in science fiction, and *Star Trek* in particular. This was the beginning of what has become known as Comic Con with cosplay—dress as your favorite superhero then attend a convention with like minds. Bjo and John Trimble were the architects. In the realm of science fiction power couples, the Trimbles were the gold standard, so there was no room for me to complain about my failed Oscar dress. I needed to be a good sport and move forward.

A Mrs. Adam called from Pacific Telephone in San Francisco to tell me that they were adding a listing for holography to the yellow pages. This was a huge victory since the phone company had initially refused to acknowledge its existence.

> Sunday, March 7, 1974 Shangri-La Ranch
>
> Met Brett, a professional mime. Went to the ranch for a meeting with Gertrude and Gerson Marks, Ed Penney, and Peter Goodgold. Went through ambivalent, insecure "feel like fifteen" syndrome with Zar. Finally realized through discussion that it is the time to act on one's intuition, not to compromise, not to make excuses for myself or anyone else. Zar emphasized, "The right combinations will happen! DO NOT FEAR!" Took a sauna. Got orders for Lumilar. Zar— "Lumilar is a gift and must be honored as such!" If I treat it with respect, all will be right.

I know the many people that overpopulate this rich hippie landscape can be confusing at times. Zar was one of the leaders of the Laguna Beach Rainbow tribe. He was a magnetic, alpha dog with a self-assured, powerful presence. His words of wisdom were true. Holograms are magical. I know, because I've traveled the world showing them to strangers and the reaction is always the same: Their eyes widen with wonder, their lips part, and big smiles form across their faces.

Obviously, some of this magic had enveloped visitors to Shangri-La Ranch in Malibu. I saw the Laguna Rainbow painters as centered and purposeful, and I wanted to attain those sensibilities. There was something about the sea-view property and salt air that invited artists to open themselves up and plug into the creative ether.

April 8, 1974

Delta Flight 41 arrives 9:56 AM. Cecile with eight bags with "Make printed holograms" on the sides. Meter maid insisted on giving me a ticket for parking too long when I had no way of carrying bags faster. Cecile's trip to Dallas exposition good. Trying to get publicity. 11 AM—all Sun Ra holograms to Meta Noia with chains for necklaces.

The Sun Ra holograms were small and on a continuous film roll. They had been commissioned by HCCA and shot at McDonnell Douglas. Along with the Ysan dragons, they were high quality and ready to be made into jewelry. Meta Noia was doing those honors.

Wednesday, April 10, 1974

Pick up Lumilar at United Airlines—lot Baltimore 4:30—pick up at Lost and Found 30 minutes after flight arrives. Call Jeff Allen—Mike Foster arrives with printed holograms— 874-0601. Jamie Scott—Bob Margoloff—Creed. Call 4:30.

Cecile had flown in for Equicon and to establish a school of holography. Gertrude Marks and Ed Penny were going to film the Jupiter 5 launch at the opening of Equicon. Cecile, Zar, and I were counting the minutes until the doors of the Los Angeles Airport Marriott flew open and hundreds of Trekkies would rush inside and into our white PVC dome with its myriad of three-dimensional treasures.

Thursday, April 11, 1974

Racing to get last minute supplies. Alan came to pick up Cecile—my car wouldn't start—I went with them. Met with Gertrude and Ed while Alan went on errands. Gertrude and Ed decide not to shoot. Can reconstruct the dome anywhere and hire 50 extras. Stayed up all night—caught an hour nap. Meta Noia with crew bit off more than they could construct in one night. Went to Malibu for movie.

Friday the show was supposed to open at 10 am. I got my car back and when I arrived at the Marriott, the Jupiter 5 structure was still under construction and the news media was pacing impatiently waiting to shoot. My father had loaned me $100 with the expectation that we needed to make change for our huge launch.

Cecile was wearing one of her more space-age jumpsuits with holographic jewelry and die cuts. I wore a long, satin A-line skirt with Lumilar die cuts sewn on, and a free-flowing blouse with lots of Jupiter 5 jewelry.

The doors didn't open until 2 pm, which meant that we couldn't sell anything until then. I took in $25 for opening day and was ready to commit harakiri. My agent, Crayton Smith, among other invited guests, couldn't get in and left. Brett Barton, the mime, a friend of the Laguna group, arrived wearing a leotard and white face. I added a few Lumilar accents and he started performing. Lumilar. I had finally invented a word that stuck in peoples' minds. I couldn't help but appreciate the *Star Trek* enthusiasts with Spock's pointy ears and Kirk's yellow pullover with black trim. This, of course, was the onset of cosplay, and Bjo Trimble's costuming handiwork was in evidence.

The Trekkies ogled our holographic jewelry displayed beneath glass bubbles and set on black velvet. They watched Zar's laser show and a few of them bought Chris Outwater's book on how to make a hologram.

Saturday was a much better day. We sold $700 worth of bracelets, necklaces, Lumilar stick-ons, and magic wands. It brought together all the people we'd been trying to convince that Jupiter 5 was at the cutting edge of fashion, and hopefully, a new industry.

Fred Butler brought comic Reynaldo Rey, who'd been selling the Ysan dragon rings at the end of his nightclub act. Ed Auswachs brought investors. Bonnie Corey came to help and stayed to party. Brett, the mime who'd recently returned from Maui, brought a powerful souvenir to share.

At the end of the day Brett, Bonnie, Cecile, and I went outside and got very stoned. "Easter is eassyyy," became our mantra. Aside from the Maui Wowie, we were beyond excited by the reception we were getting at Equicon. Easter was the next day and we were primed to celebrate. And celebrate we did.

We ran into Mr. and Mrs. Flash Gordon, aka Tallie and Patrick, who invited us to a party in a hospitality room. Clearly, they knew their way around and a locked door didn't dissuade them. We watched as the costumed superheroes instructed a young hotel employee to unlock the door. Obviously, we were exhibitors and convention organizers. Wink. Wink. The suite was spacious with a full bar, music and comfortable couches. Everything we needed for a fabulous complimentary night at the Marriott.

April 12, 1974

Cecile and I settle into working on the show. We've told everyone in the news media, that we're going to have an exhibition of holograms, including some of Salvador Dalí's and Dr. Wuerker's (TRW). I invite a long list of interested people.

The show is a nightmare. The fireproof dome cost $750. Jupiter 5 owns half of a fireproof dome that it takes six men two days to erect...

The show is about to open. Nick 'Mission Impossible' Franzosa isn't ready. I've had two hours sleep. Alan's bought the wrong items which I have to exchange, and we don't have the Kirilian test unit anywhere near ready. No holograms are up, there is a slight laser show, and we have Tubular Bells playing in the background.

The news media comes and goes because they can't shoot an

unfinished exhibition. Cecile gets a great blurb in the Daily Breeze. Bret Barton performs mime, and I calm a bit. Gertrude Marks comes down to see if this is worth filming...after the first half-assed day the answer is, no.

Scotty arrives with Donyale Luna and a whole new world opens up.

I knew Scotty, a tall, very handsome black actor and model. I had invited him to the show, and in turn, he had brought Donyale Luna. Donyale was one of the most exotic and beautiful creatures on the planet. A willowy six-feet-two-inches with creamy milk chocolate skin and patrician features, she moved with the allure and assurance of an otherworldly queen. In 1966 she was the first black model to grace the cover of *Vogue*, and on April 1, 1966, *Time* magazine had heralded 1966 as "The Luna Year." This was a revolutionary acknowledgement given that the first time she appeared on the pages of *Harper's Bazaar* in 1965, a number of advertisers withdrew, and many Southern subscribers cancelled their subscriptions. Racism was proving to be an uphill battle in America, so Luna moved to London and instantly became Europe's most sought-after cover model. The *Time* magazine article describes, "a new heavenly body who, because of her striking singularity, promises to remain on high for many a season.

Skyrocketing to superstardom in London during the Swinging Sixties introduced Luna to a glamorous lifestyle that more often than not included drugs and alcohol. Federico Fellini adored her and cast her in his 1969 film, *Fellini Satyricon*. Salvador Dalí invited her into his inner circle, and Andy Warhol captured her remarkable essence in three films at the Factory.

Luna, as most people called her, had been born Peggy Ann Freeman in Detroit, Michigan. She was a middle-class black girl, tall, gracefully thin, with a penchant for making others believe she'd landed from a galaxy far, far away.

Dressed in regal robes, their arrival at our show prompted a parting of the human sea. Costumed gawkers moved out of their way, watching, and wondering who these two beautiful, otherworldly figures were. As the duo began to ogle and comment on the three-dimensional images draped on black velvet and floating beneath Plexiglass bubbles, our exhibit became the convention hotspot.

Luna bought a Ysan Dragon ring, urging those within earshot to secure a piece of the future for themselves. And they did. We sold magic wands and packages of stick-ons, holographic necklaces, bracelets, and rings. It was as if the king and queen of the universe had blessed our wares.

I miss Luna. She wanted me to come to Rome to meet Fellini. "Fellini will love you and your script," she would tell me in her low whisper. And I believed her. There was nothing I would have rather done, no place I would rather have traveled. I'd minored in Italian in college, and at that time I think I was still conversant.

> Sunday, April 14, 1974
>
> Want to make money. Work constantly, no sleep, high on event! Scotty (Jamie Scott) brings the most beautiful woman in the world, Donyale Luna. She's the prototype for Galina, one of Fellini's favorites in Satyricon. Shangri-La—meeting of space angels for Bob Brown's self-contained electronic musical journey. Scotty and Donyale arrived—all were equally awed by their inner and outer beauty. Donyale is a true queen.

Luna, as she was called, and I became close friends. She was that rare flower that one wants to protect. She was perpetually late. A frustrating friend, but one with an open, all-embracing heart. She flipped over the holographic jewelry and it looked spectacular on her.

> Monday, April 15, 1974
>
> Woke up at Shangri-La. Drove to office. Called Luna. Went to Old World for dinner. Quick cause Luna in a hurry. She orders—knows what's best—champagne with orange juice and salad. She wants three candles and they'll only give her one. "These humans have no appreciation for the divine."

Tuesday I met Donyale at the Golden Temple, a vegetarian hub on Sunset. "I have told Mr. Freddie Fields all about you," she whispered in her breathy voice. Beautiful Donyale had an entourage of artists who all loved her. When she spoke she seemed to pull each word from the ionosphere. Long pauses punctuated every sentence, forcing the listener to lean in and hang on every syllable.

The Equicon convention had spawned a whole new, wonderful group of connections. It felt like I'd boarded a bullet train to the next dimension.

> Wednesday, April 17, 1974
>
> Gertrude Marks—noon. Got up late, Cecile taking phone calls. Selling Lumilar. Donyale called, "Freddie Fields will call...wait." Bonnie (Corey) arrives with our clean laundry. No Mr. Fields. Take a nap. Mr. Fields calls. I'm in bed wearing hot rollers and cooing holography, the magic art of lasers, Lumilar.

Freddie Fields was a legendary agent who handled the biggest stars in Hollywood. As a co-founder of CMA (Creative Management Agency), which evolved into ICM (International Creative Management), he was the ultimate kingmaker.

"How's eight-forty-five?...no, nine o'clock?" Fields asked.

"Fine! I'll pick up Donyale."

"Oh, no," he shot back, "she's not coming. She understands...the lady who's living with me now doesn't understand Donyale. You understand?!"

Whether I understood or not didn't matter. I apologized to Donyale and Cecile and I put together a collection of holograms and Lumilar samples. We wore our Jupiter 5 jewelry so the fashion spoke for itself. Fields' house was more mansion than house. His girlfriend Cheri was cut from the same cloth as today's *The Real Housewives of Beverly Hills*. Every hair was in place, and I would say platinum was more to her liking than plastic covered in Lumilar.

A houseman ushered us into Freddie's full-sized screening room, a spacious theater that could comfortably seat one hundred guests. Drinks and grass strong enough to please a Rastafarian arrived, and led to a discussion about my *Revelation II* script and the larger-than-life-sized special effect necessary for first run theaters.

"I want to tie up the Diffraction Company for a-hundred-and-twenty days," Fields proclaimed. "And the material needs to be wider!"

Freddie Fields' enthusiasm was contagious. I knew if he wanted to package *Revelation II*, my film would get made. He and his much younger girlfriend were blown away by the three-dimensional images and diffraction grating samples. That night Cecile and I floated out the door. It wasn't until we got back to my apartment that we realized that we'd left the holograms in the screening room. Of course, the following morning, embarrassing as it was, we had to make a return trip to collect them.

Thursday Donyale wanted me to take her to buy a new suitcase. On the face of it, a simple task became an exotic foray into Robinson's, a conservative Beverly Hills department store, with Cecile, Bonnie Corey, and Kaisik Wong, a designer friend of Luna's who Salvador Dalí regarded so highly he gave him a room in his museum in Cadaques, Spain.

Kaisik, not unlike Luna, was an original. His work celebrated his dynastic Chinese heritage with luscious hand-embroidered silks and satins. He cut his own patterns and sewed his creations by hand. Kaisik Wong is credited with pioneering wearable art.

We all piled into my white Toronado with Lumilar lightening bolts flaring in the sunlight and made our way to Robinson's. Between Luna's great height and soft airy voice, and Kai's ebullient personality, shoppers gawked, some wondering who we were, while others broadcast disapproving stares. I was in heaven. I was broke, but I felt rich.

April 19, 1974

Greatest night I can remember. Arrived at Kaisik's to find Richard preparing dinner, Kathleen serving champagne. Large home overlooking L.A.—Gerhardt—composer. Steppi, star of Stephen Arnold's Luminous Procurus. Thought we, Cecile, Bonnie and I were three hours late to dinner party, and actually we were three hours early. Stopped at Chateau (Marmont) on the way and bumped into Freddie Fields as we rounded a corner

> to the elevator. Luna has a crystal ball on one pillow and two old broken dolls on the other (from childhood). She keeps goodies in stuffed rabbit. So beautiful is she who controls this small elite universe. Freddie Fields was very happy to see us. The power of optimism. Kaisik showed us his line. The most beautiful oriental handiwork. "I only want to dress for the court," he told us.

Sunday I drove Luna to the airport, making sure that her many bags, fur coats, jewelry, and feathers made it onto the flight to New York where she would be joining Salvador Dalí's court at the St. Regis Hotel.

CHAPTER 17

MEN WITH FLASH TARNISH FAST

Monday, April 22, 1974
3 pm Westwood Chamber of Commerce. 4:30 Ed Penny. 6 pm Jeff Allen 2027 Curson. Kaisik (Wong) called, "I'm going to New York to see Dalí." Kai took the letter that Cecile dictated—flew to NY. I couldn't reach Andy (Pfeffer)—finally got correct legal letter for Dalí to sign from Andy's secretary. Stopped at Reynaldo's who was angel dusted out and called Kai at Pandora's just before he met Dalí for dinner. Perfect timing. Met Jeff Allen who discussed Mike Foster's production possibilities and their deal with Mattel to make toys you can see around sides—different views from different angles (3-D).

Being swept into the Salvador Dalí circle made for intensely exciting times. I was working on a holographic movie script starring Salvador Dalí and Donyale Luna. It began with Dalí appearing in a Lumilar robe and crown. He would perform magic tricks and Luna would become his creation, the high priestess. He transforms her into a life-size hologram, then creates Kaisik, and then Steppi (Steven Solberg), the court jester. A coronation ceremony unfolds, with Dalí being crowned king of the universe. The remainder of the film would depend upon Dalí's wishes.

On Tuesday, Cecile and I spent two hours being interviewed by Beth Ann Krier for an article in the *Los Angeles Times*. By the time we left, a man from the photography department wanted to buy enough Lumilar to make a backdrop.

It was during this period that I began spending a lot of time at Bob Margouleff's beach house. His assistant, Creed Yankovich, was twenty-something with a golden tan and a wide, milk-white smile. Both men and women were magnetized by his Rhett Butler-esque charm.

Working for Bob Margouleff, one of the busiest, most in-demand recording engineers and producers in the music business, gave Creed ample time to host many memorable Malibu beach parties. It was at these gatherings that I developed a ready market within the music business for Jupiter 5 holographic jewelry and Lumilar. In many ways, I was the flavor of the month, and as such I got to party and sell my wares simultaneously. It was not unusual to see Stevie Wonder or Minnie Riperton with her young daughter, Maya Rudolph, balanced on her husband's shoulders. I happily found myself at the intersection of business and pleasure.

> Saturday, April 27, 1974
>
> Went to Malibu (Margouleff's) to see Creed and have letter signed requesting quote for 250,000 holograms. Took Beth Ann Krier to experience one of Creed's parties.

I was flying high when I received a call from Mi Flick. "Remember what I told you," she urged. "I'm sending you someone from Seattle. He's very smart, in the music business, and he's moving to L.A."

I trusted Mi. I'd met Bob Flick, and his folk group, The Brothers Four, was successful all over the world. It made sense that Mi would spend time with music people in Seattle. "His name is Richard Aaron. Take him seriously."

Ironically, there were two Richards in the music business with very similar sounding last names. There was forty-year-old white Richard Arons, who was co-managing the Jackson 5 with Joe Jackson, and black

Richard Aaron, the twenty-seven-year-old Harvard graduate whom I was told was baseball legend Hank Aaron's nephew. For young Richard Aaron, this coincidence could be used to his advantage. Tall and slender with what one might describe as a runner's body, the young talent manager and budding music executive hit the hot asphalt running. The first time I met him, he was wearing a suit and tie. He had boundless energy and a genuine passion for music.

I wanted to be swept off my feet by a dynamo, a man who would throw doors open and let me work in peace. Part of this was generational and part was the climate of the seventies. It was very much a man's world, and I needed one with talent and ambition who shared my sensibilities and vision for the future. I thought I had found that man in the person of Richard Aaron.

Our relationship escalated quickly. Richard didn't have a car, so I drove him to various record companies and recording studios. He was looking for songs to place with groups from Motown, Stax, and Warner Brothers, while I wanted to sell record executives on holographic album covers and store displays. We complemented each other. At least that was how I rationalized spending so much time at clubs and record companies.

One day Richard took me to meet jazz pianist Walter Bishop, Jr. He told me that Walter had played with Charlie "Bird" Parker. "If you could do anything right now what would it be?" Bishop asked. "Write?"

"Yes," I affirmed, and he offered to pay me $7,800 to write a film treatment for a feature about the life of Charlie Parker. That was music to my ears.

Richard was always in motion, making appointments with a variety of executives in a variety of cities. I saw him in action and I was impressed with his commitment to the performers he was representing.

"If Fred Butler's tunes are as good as you keep telling me, I can get him a good contract and two hundred a week."

Motivated by either guilt or friendship, I arranged a meeting between the two music men. Fred played a couple of his songs for Richard, who responded positively. He said he was looking for tunes for a gospel-turned-pop group called the Emotions.

What I thought was a good deed definitely did not go unpunished. Fred picked up on the chemistry between Richard and me and the meeting ended. Fred was furious, insisting that Richard Aaron was a complete con artist. A hustler! Of course, in my mind Fred was behaving like a jealous, jilted boyfriend.

There was no time to stew over romance gone awry. The Westwood Art Fair promised to be a big money maker. Westwood was an affluent college town bordered by much wealthier Bel Air, and I believed that both UCLA students and locals would find Jupiter 5 jewelry and holograms irresistible.

Beth Ann Krier's article on Cecile and me was published in the *Los Angeles Times* along with pictures of us, the futuristic duo. I found it thrilling. The piece was positive and pointed to us as entrepreneurs focused on innovating art and fashion.

> Sunday, May 5, 1974
>
> Westwood art fair. Overcast. Bad lighting for holograms. Sunday no better than Saturday. Seeing Times article good for p.r. Finally receiving public recognition for something. No word from Richard Aaron, no money, great press, we need a monetary miracle to pull us through. Zar too dogmatic. Says God motivates him. He knows my weak spots and wastes no time puncturing them. Cecile finally realized that (my) parents are totally unimpressed by the (L.A. Times) article. They think I should be making money—good reviews don't mean anything.

I discovered this description of the art fair in my diary and it really captured the day:

> Jupiter 5 paid its $35 and got in under the wire. We moved towards what we or at least I thought would be a "monster" sidewalk show. Unfortunately, the sun refused to shine and the ambient conditions made the jewelry look flat, dull. It was amazing. Our picture was in the Sunday Times, a major article

scheduled for the first page of View...pushed back to page 22. No matter, the article read like a press piece making Cecile and me super women. We do our best and manage to sell under $100 worth of jewelry and magic wands plus get ripped off for God knows how much. It was another nightmare. The only time the sun came out there was a gas leak in the building next to us and five booths had to move. We were one of them.

We pile tables and hopes and another expenditure of energy into the one lightning bolted car and hobble back to the motel-apartment. Then the phone starts ringing off the hook because of the Times article. People want information on holography. They want to see one, they want a brochure, they want to invest...

On Tuesday, May 7, Cecile and I were interviewed on KABC by Elliot Mintz for his talk show. Word was out that we were up to something extraordinary. Donald Camell, who had co-directed *Performance*, the cult classic starring Mick Jagger and James Fox, wanted to speak to me about directing the Dalí film.

I was covering Hollywood while Cecile was embedded in Laguna with the Rainbow people. Richard had traveled to Memphis where he heard a young white singer/songwriter with a throaty, soulful Janis Joplin sound. Her name was Julia Chuhralya, and he brought her to Los Angeles to get her a record deal and make her a superstar. It became my job to drive them to record companies all over town. At the same time Cecile was working on setting up a School of Holography in Santa Monica.

Wednesday, May 8, 1974

10:30 am Rich Bobrick (potential investor)—Hyatt House. Bobrick: "I know I could borrow $40,000 or 50,000 from the bank, but I just don't think I could borrow enough to factor you on the 15th of every month. I really think this is going to be a $50 to $200,000 a month business." The man was shaking. "I'm not sure without test marketing, but that's what I feel." Life is moving ahead as fast as acceleration will allow. Learned you can iron Lumilar.

People on the outside saw us as influencers and disruptors. What they didn't know was that behind the scenes we were too caught up in the heat of the moment to make sound business decisions. I got a call from Jody Burns in New York. His usual cool was gone. Concern tempered his words as he explained that someone had stolen the projector and holographic film. Donyale Luna was MIA and he asked if I had any idea where he could find her. Luigi, Luna's Italian boyfriend, was taking a course at the School of Holography in Manhattan, and the Italian director didn't understand why he had to pay for the privilege.

"Much as I'd like to help you, Jody, we just did a show and the dancers walked off with twenty-seven Lumilar bracelets." New York would have to deal with New York's problems. I had plenty to deal with on the West Coast.

Monday, May 13, 1974

Richard (Aaron) mentioning he's late for marketing ethnic films in London. Waiting for his Seattle bank to transfer his money. Richard says he's placing tunes with Roberta Flack, Jackson 5, Pointer Sisters—calling all over the world. Cecile and I saw the nationwide Bicentennial celebrations as a way of introducing the public to holograms. We believed that holograms could transform boring two-dimensional exhibits and provide a unique opportunity to produce souvenirs commemorating two hundred years of independence.

We rehearsed our pitch as we drove downtown to City Hall. We needed funding and Cecile and I believed that the American Revolution Bicentennial Commission had the power to connect us with the right financial sources.

We showed the men holograms, and they were very receptive. Anyone could see that three-dimensional imagery done well would leave a lasting impression. Holograms would make the Bicentennial more memorable. There were a number of categories and it was suggested that we apply to the Council for the Arts for funding. People wanted 3-D imagery, but they wanted us to find the money to produce it. They weren't saying, "Call my contact at Bank of America or Wells Fargo or TRW." It was more like, "We'd love to have holograms! You girls are smart—make it happen!"

Raising money was challenging. However, since landing the shoe page in the *Los Angeles Times* Sunday magazine, Fred Slatten had been sending Jupiter 5 Lumilar art commissions. Dennis Pelletier, aka Amir Façade, did Elton John's boots and the Jackson 5's shoes.

Percussionist Rocky Dzidzornu, a Ghanian-born English musician known for playing with the Rolling Stones, Taj Mahal, and Stevie Wonder, was having us transform his conga drums using five different diffraction grating patterns and colors. Dennis cut birds and bugs in a way that made them kinetic when the stage lights came up.

Wednesday, May 15, 1974

Incredible day. Dashed to Malibu to see townhouse. Nice but not funky. Called to see two bedroom down the beach. All talk about Fred Williamson's house and $1,300 per month faded as we were taken to a house near Creed. Burglar alarm went off—great, tastefully furnished. Said we'd take it. Gave a $200 deposit. On to Bill MacDonald's Sunday soiree for 300 hip, "in," making it L.A. youth types. The young and the beautiful on top of Sunset Plaza Drive overlooking everything! Top of the world! Saw and confronted Sandi Sircus who ran the Paradise Ballroom for Jerry Brandt. She told me, "Six months ago I'd a punched you. Cary was my old man—Jerry never liked him, and he brought you in. I've been told you were naïve, but nobody's THAT naïve."

I was that naïve. I grew up in a bubble, a luxurious bubble. The people around me had all the trappings of success so I supposed that, by proximity, I too would be successful. Novelist Sidney Sheldon lived next door and "Mommie Dearest" Joan Crawford lived down the street. I took people at their word, and when Richard Aaron started complaining about his bank in Seattle, I wasn't surprised or alarmed.

Every day he shared his mounting frustration. "They're making me jump through hoops. Stupid red tape! My money should've been transferred two weeks ago!" I couldn't help but feel empathetic because I saw how hard Richard was working. We were both running to meetings and checking out rental properties, and at the end of the day, heading to clubs and parties.

The parties were as much business and networking as having a good time. Cecile often stayed in Laguna with Zar while I put on hologram and jewelry shows in and around L.A. Julia would play her guitar and sing while Richard might nod off from exhaustion. Every project he took on was, in his words, a "winner." He expended a tremendous amount of energy, but so far nothing had catalyzed. Richard wasn't the only one dealing with frustrating circumstances. I discovered that Zar had borrowed two lasers that no longer worked, plus Cecile had borrowed holograms from Dr. Ralph Wuerker at TRW and a New World Space Man hologram and light source from McDonnell Douglas. If these things weren't returned immediately, my relationship with these key corporations would be ruined.

And to top everything off, to make sure my eviction was complete, my landlady left me a note saying that I had to be out by Monday!

May 23, 1974

Cecile supposed to leave—take her and all the bags to the airport. She's been up all night packing—picking up apartment. Drop her and leave. I'm relieved. For all the good she's done there's an equal amount of confusion and fuck ups. Losing all the bracelets, my good skirt, giving Larry Spain my mannequin, painting, jewelry, etc. Never asking me, running up immense phone bills, etc. So whappo—she calls—Jody (Burns) hasn't prepaid ticket and she's stuck. Richard takes us to Jackson 5 party. I'm so thrashed I could care less.

Every time I saw Zar and his backup chorus, Silverbird, I heard the same reprimand, "You're a Sagg (Saggitarius). You're aiming your arrow in front of you, not up. Look to the sky! Shoot high!" The message was that I was missing the true meaning of Lumilar. That I should be treating it more reverentially.

I didn't care if I ever saw the celestial duo again. Zar said a line and Silverbird reinforced it. I felt that they had thrown Cecile, my friend and business partner, completely off course. The rift between us kept widening.

Saturday, May 25, 1974

Bizarre moving day. Richard and I asleep—door latches bolted. 11:20 Mrs. Mio appears and demands that I get out by noon. Richard, Julia and I move all of my belongings to my parents' house. All exhausted. Julia takes a nap. Richard goes to his sister's to change clothes and returns at 10:30 pm—I'm nearly hysterical because I don't know where he is. Julia's at the Holiday Inn.

It was a holiday weekend and Sunday was spent at Bob Margouleff's Malibu beach house meeting the talented cast of *The Rocky Horror Picture Show*. I now had dichromate holograms, magic wands, Lumilar stick-ons, and jewelry. Dichromates, unlike other holograms, floated in a golden bath of pale turquoise, mysteriously appearing, magically popping and changing 3-D angles as the light shifted. Bob Margouleff borrowed a printed hologram on black vinyl to send to Germany for consideration as a record cover. I sold enough to take us to Zucky's, a deli in Santa Monica, for dinner. We all got indigestion, ending up at the Holiday Inn with Julia singing us to sleep. Word of Donyale Luna possibly appearing on the Christmas cover of *Playboy* wearing Lumilar bracelets and surrounded by artfully integrated yards of diffraction grating patterns should have been great news. And, in the beginning, it was. Jody Burns, the sharp New Yorker, fell under Luna's intoxicating spell and ordered one full roll of every pattern and color of Lumilar from the Diffraction Grating Company.

Seven thousand dollars worth of product was being shipped to HCCA in New York—but without a purchase order from *Playboy*.

It was the perfect storm. Luna believed that she was going to be on *Playboy's* Christmas cover, and I have no doubt that Hugh Hefner discussed the possibility with her, but nothing was set in stone. In the end, Luna only posed for an inside spread. Jupiter 5, however, got the bill for $7,000.

I called my agent to let him know I needed a writing assignment. The following week he set up a pitch meeting with a producer for a sitcom. The producer didn't like any of my ideas, but he insisted on buying $75 worth of the holographic jewelry I was wearing. I kept telling him that I was really a writer, not a businesswoman. I asked myself why these two endeavors should be mutually exclusive and I couldn't come up with one good reason.

> Monday, May 27, 1974—Memorial Day
>
> Needed to be alone. Left Holiday Inn and came to quiet of Brentwood. We have $5 between us and each of us is nerve shattered. Julia's going for a walk. Richard's staying in bed all day. I feel like running away only I can't. There's no place to run since I must move and clear up the NY $7,000 debacle with Playboy.

Now we've come to the most gut-wrenching part of the story. I mentioned that Richard Aaron was having trouble getting his funds transferred from Seattle. This was putting pressure on me to pay for food and gas, which appeared to fuel his frustration even more. Finally he found a solution. "Cash a personal check for me and in a few days when I have my money, I'll give you cash."

At first I hesitated, then I decided I was being too uptight. I was certain that his funds would arrive any day and he would pay me back.

I took his check for $750 to my bank, cashed it and gave him the money. We carried on, business as usual for three weeks until I received my bank statement.

When I opened it and discovered that all the checks I'd written to vendors had bounced, I became hysterical. How could this be?! I diligently accounted for money in and out. Bookkeeping was my least favorite task, but I did it.

Upon further investigation I discovered that Richard Aaron had done what's known as check kiting. You give someone a check with the expectation that you'll be able to put funds into your account before the bank processes it. Richard thought one of his deals would pay off before his Seattle bank processed his check.

Not only was I in the hole for my vendors' payments, but the bank was charging me an overdraft fee on every bounced check. I fought with the bank—how had it taken them three weeks to notify me of Richard Aaron's insufficient funds?!? They finally dropped the overdraft fees.

There was no mistake. As Fred Butler had warned, Richard Aaron was a con man. Now, not only did I have a con man for a boyfriend, but a boyfriend who had caused me to jeopardize my relationship with vendors I thought I was paying on time.

I knew I couldn't go to my parents for assistance. They'd tell me how stupid I was and in this case I'd have to agree with them. Instead I went to Shelly Davis. His years at the Whiskey A Go Go, gambling in Las Vegas, and knowing how to deal with disreputable people, made me think he could tell me what to do. And he did.

"Tell the bastard to give you your money! Keep after him—make him get it from somebody else! If you keep after him, he'll get you your money!"

So I kept after him and it wasn't easy because he assiduously avoided me. Ultimately I had to fill out a formal, notarized complaint. In it I stated that I had called him dozens of times and gone to see him dozens of times, met him at 8899 Beverly Blvd. in Bobby Roberts' office to collect money that I was told he was receiving, and called Richard Arons of the Jackson 5 to verify a supposed commission.

Shelly Davis believed that Richard had done this before. He was calculating, and if he was as sharp as I had originally thought, he'd be able to get the money from somebody else and repay me. "Just keep after him, Linda! Keep after him!"

CHAPTER 18

SORTING THEM OUT

Anticipation, confusion, and crisis heightened every waking, sleepless, couch-surfing minute. Mi Flick was in London, too far away to help me with Richard Aaron. I was stuck with financial dyspepsia and a June 30th deadline for Disney's *Herbie Rides Again* VW bug contest. If our Lumilar Bicentennial-themed car won, we'd hit the PR jackpot and drive home in a brand-new Volkswagen.

Monday, June 3, 1974

Picked Selwyn up at the airport, went to office, then Ankrum Gallery. He met owners and was blown away by Shirl Goedike's beautiful work. A contemporary master! Shirl's lucky number is "5." He was taken back by 555 N. Bristol and Jupiter 5. Selwyn set up new holograms (Ankrum Gallery)—Donyale (two shots) wearing Lumilar, Sam Rivers (best), Selwyn's art piece—girl looking in the mirror—not dynamic enough—holograms received as medium of the future and not ready for exhibition. Mary Goedike (Shirl's wife) bought eight feet of star pattern and 1" gold and 1/4" mosaic to take back to France. I went back to office and sold $200 Lumilar so my Jupiter 5 account will be okay.

> Always living on Lumilar. Selwyn always turns negative. Says he's selling HCCA stock for $15,000.

Selwyn Lissack and I had a somewhat contentious relationship as is evidenced from the following archival writing.

> ...how funny...how bizarre...or is it bizarre?...one doesn't know anymore just what is true and what is pure fabrication. Each individual has "A Side" and "Each Star" has their best side, that pose that suits their image of themselves the Best!
>
> That's where we are today in the world of HOLOGRAPHY. It's oh so bizarre to listen to Selwyn describe the multiplex process as his very own "exclusive" process. "There's one camera in the world..." "Is there any competition?" "...ah, no." And he shakes his head in a low profile manner allowing everyone to stare at the large brown-eyed man with hunched shoulders and a peeling sun tan, silently sighing, "What a genius."
>
> Of course, that is fair. Selwyn probably is a contemporary genius. That part is well and good. It's the other side of the coin that makes me wince.
>
> We're in Bob Banner's office. It's a Saturday. Steve Pouliot is a young man with a brilliant future. He sees the potential in the portrait setup and says, "Tell me what I can do to help you?"
>
> Selwyn shakes his head and suggests a series of "star portraits," which is a logical suggestion. He never once mentions the fact that he's affiliated with HCCA or with me who happens to be a writer. He plays the whole scene humble and literally comes off as best friend of Salvador Dalí and creator of this magical new four-dimensional process...then we digress. What has made me so irritable? Why have I no place to live? Why am I staying one

night in Malibu and one night in Hollywood? Why am I living out of a suitcase?

Networking came naturally to me, so I was the door opener. I've realized that my business associates took that ability for granted. Once they were in the room, I became expendable. Bob Banner was a legendary writer, producer, and director in the Golden Age of television. He'd produced *The Carol Burnett Show* from 1967 to 1972 and won a number of awards for *The Dinah Shore Chevy Show*. Steve Pouliot was a young writer working for Bob Banner Productions. He saw the potential in what we were doing and wanted to help us.

Unfortunately, Selwyn was too short-sighted to see that L.A. was a small town, a company town, a place where you couldn't hide the truth for long. The blue Pacific played a siren song, drawing holographers and their handlers to the West Coast, each vying for a slice of the 3-D market. I rationalized my deep dive into the HCCA/Jupiter 5 business by convincing myself that the person or company capable of making the larger-than-life-size hologram I needed for *Revelation II* was just around the corner, about to cross my path.

So far I'd met two types of holographers: the academic physicists who wanted to party with rock stars, and the more grounded ones who saw themselves as pioneers in a long-term nuts-and-bolts business .

Based on the popularity of Lumilar and the products we were making, I wanted to open a Jupiter 5 store. I envisioned a gallery space where different types of holograms could be exhibited and Lumilar could be sold by the foot. I felt sure that bringing all things holographic together in one space would give Selwyn, Cecile, and me a chance to build a solid bicoastal base. Only I wasn't the only one with this idea. Burton Holmes International on the Sunset Strip desperately needed to find a new source of revenue, and holography was a natural, potentially lucrative choice.

Burton Holmes had coined the term *travelogue*, and when he died in 1958, he left behind miles of film and photos dating back to the First World War. By 1974 the audience for travel lectures had all but disappeared, while holograms were just taking off. By setting up a multiplex studio in the heart of Hollywood, adjacent to Beverly Hills, Lloyd Cross and Burton Holmes International were staking their claim.

Bob Mallett, the head of Burton Holmes International, was a charming older gentleman who had worked for the famous Mr. Holmes and eventually had taken over as the company's travelogue lecturer. When I met him he was tasked with keeping the prestigious company afloat.

At Mallett's behest and in his absence, Bob Hollingsworth, a dynamic six-foot-four-inch, 350-pound engineer, inventor, and jack-of-all-trades, took charge. Hollingsworth had started working for Holmes when he was a teenager, and even though he'd married and had children, BHI dominated his life. He was driven to make the company solvent again and his gut told him that holograms were the means to that end.

In the beginning what no one knew was that Burton Holmes needed financial success as much as the holographers themselves. The rent was usually late or in arrears, and were it not for a generous landlady who adored the company's eponymous founder they would've already been evicted.

Bob Christiansen was my first contact with Burton Holmes International. He was tall and gaunt, older than I was, and hungry for a commission. Sometimes he just looked hungry. Everything was in its nascent stages. Lloyd Cross came down from San Francisco and set up a multiplex studio at BHI. And John Foy, a ruddy faced, red-headed jack Mormon from Utah, came in with diffraction grating trip glasses that he said had been produced in a secret Salt Lake City basement.

Salt Lake City was a nine- or ten-hour drive from Los Angeles, and holographers saw the entertainment capital and government-funded aerospace industries as their target. Word was that Michael Foster was the genius of the new medium. That he had invented something Shell Oil had patented, making millions for them and only a chemist's salary for himself.

Foster was an outlier. The mere mention of his name brought reverential sighs. In those days everyone was skinny. Foster was skinnier. He and his beautiful Botticelli-blond girlfriend, Randi, made business trips to L.A. to meet with record company executives and toy manufacturers, and basically anyone willing to belly up to the table with big bucks for a prototype. Michael Foster was the real deal and I believed that if anyone could make a larger-than-life-size hologram, he was the one.

Getting an introduction however was impossible unless you went through Jeff Allen, Foster's Salt Lake City partner and front man. Allen was tough, with dollar signs shining through his aviator sunglasses. In private, we referred to him as "the snake."

Foster was the grand prize, and anyone who knew him was sought out. John Foy, a partner in Lumens, Foster's first corporation, adopted Burton Holmes as his temporary office, and eventually Lumens' other partners, the Wanlass brothers, Jim and Mike, relocated to Los Angeles to complete the genius' posse.

John Foy was selling Trip Glasses using Michael Foster's clear diffraction grating material while I had formed Optical Infinities with Lloyd Cross to manufacture and market the same product. The patent attorney had warned me about this. If we couldn't patent anything, we needed to get our glasses out first, and Lloyd was on to other things, and I was focused on Lumilar. Trip Glasses were John Foy's lifeblood.

BMI on the Sunset Strip was quickly becoming the hub for holographers. Richard Rallison, another Salt Lake City transplant, had landed a job making holograms at Hughes Aircraft in Culver City. By day he shot industrial images and by night he toiled over dichromate emulsion. Of course, his dream was to have his own fulltime holography business.

Dichromate holograms were the easiest to see and, you might say, the prettiest, with their greenish blue and golden hues. The tricky part was making your own emulsion. When it went well, the images were clear, crisp, and mysteriously magical. If the emulsion wasn't perfect, there might be a partial image or nothing at all. And, finally, if the plate or bottle wasn't airtight, chances were the image would attract moisture and disappear altogether.

Richard Rallison knew Michael Foster from Salt Lake City and credited him with teaching him valuable shortcuts for making good, stable dichromate gelatin plates. He also credited his mentor with special developing shortcuts. Rallison started using bell jars, shooting holograms of coins, seashells, and jewelry, then sealing the lids. The result was a beautiful art piece that appeared to be filled to the brim with glittering objects. Rallison also made dichromate jewelry.

> Wednesday, June 5, 1974
>
> Selwyn said when Cecile couldn't pay rent at 865 (Broadway) anymore Jody (Burns) gave her and (her son) Craig a place to live. So many sides to every story.

Selwyn and I met Rich Rallison for dinner and found his road to holography fascinating. He confided that he'd been busted in Utah for pot possession and gone to jail for six months. That gave him an opportunity to study metaphysical books and expand his consciousness. Afterwards, he studied holography and landed a job in California at Hughes Aircraft. Although clean cut and in his early twenties, he was lucky to have chosen a field so small that his youthful blunder was overlooked.

I bought seven holograms from him for $22 and Selwyn bought hologram glasses. Suddenly we had another source of product, and Rallison was down to earth and reliable with a great sense of humor. In his approach to holography, he was more like Chris Outwater than his aforementioned Utah hippie counterparts.

> Friday, June 7, 1974
>
> Richard (Aaron) has blown all his deals including with the other Richard Arons. Group "Sundown" came in from Mt. Baldi to sign bank loan and waited all day for Richard (Aaron) who never showed. One guy actually cried. Jeffrey (Dengrove) called the Source and told them to give the group coffee and to charge it to him. Julia holding up alright.

The walls were rapidly closing in on Richard Aaron. He'd promise to meet me at the bank or at an office, then he'd call with an excuse or just not show up at all. It was a hopeless pattern, and the only way I could keep from going crazy was to sell enough holograms and diffraction grating material to keep vendors at bay.

Everybody knew somebody who needed holograms. Jeff Allen wanted to take orders for Michael Foster's printed or embossed holograms, and Selwyn wanted prices for them printed on vinyl. One person knew someone who played tennis with an heir to Max Factor cosmetics. He wanted to show him holographic diffraction gratings for Christmas packaging. Someone else was in contact with Barry Gordy and wanted him to order embossed 3-D records for a Motown group. The possibilities were limitless.

While Los Angeles holographers were jockeying for position, Lloyd Cross and the School of Holography in San Francisco had moved on to pornograms. Pam Brazier, Lloyd's love and star model, shot Pam and Helen—two naked beauties kissing—Pam falling to her knees and licking Helen's breast while the 36-inch cylinder slowly rotated. Was it art or kitsch? It didn't matter, because people had never seen anything like it and they were mesmerized.

Meanwhile, Julia Chuhralya, the singer Richard Aaron brought to Los Angeles from Memphis for a record deal, kicked him to the curb and moved in with the other Richard—Richard Arons, co-manager of the Jackson 5.

Wednesday, June 12, 1974

Richard (Aaron) back from Ohio Players supposedly with commission money. Picked up Selwyn—went to Jeff Allen's to make a deal for revised price list. I went by The Source as Richard called to say he wasn't going to be there. Meeting with Rich Rallison yielded a written agreement for six months and beyond. We buy 24 tumblers a week and he'll create original subjects.

Changed—drove in from beach with Selwyn for 12 o'clock meeting with Gary Silverman. We were late and Silverman never showed so we pressed on to Malcolm's (Malcolm Cecil's in Malibu). Waited for Bob Buffington and Philip and 360 light source. They put that together while we made Lumilar pins.

Dennis Pelletier got into ornaments using basket weave pattern in opposite directions to create a fourth dimension to a symmetrical (Xmas) tree.

With the end of the month in sight, we began wooing a VW bug owner named Mike Van Horn to loan us his car for Disney's Herbie contest. We were all certain that sunshine hitting a Lumilar art piece designed by Dennis Pelletier, aka Amir Façade, would carry us to victory. Dennis ran Malcolm and Paula Cecil's ranch style house at Point Dume. It had become my oasis, high on a hill within eyeshot of the blue Pacific, and down the street from Bob Dylan. Being there with creative, talented people allowed me to leave my worries at the door.

Malcolm, an acclaimed English jazz musician, engineer, producer, and the genius creator of TONTO, the room-sized synthesizer, was busy 24/7. This left a nice window for Malcolm's wife Paula, son Milton, Dennis, assorted friends, and me to play with Lumilar. On this particular day we made two dozen pins, which I immediately placed at a store in Hollywood.

In my diary I wrote:

I am a high wire ready to snap. Awakened by tension—what to do about Richard Aaron.

I was being consumed by an impossible situation. My producer friend, Shelly Davis, kept telling me that if I kept after the con man, he'd get the money from someone else and pay me. Well, this didn't seem to be happening. Not even a little bit.

Saturday, June 15, 1974

Bob Banner Productions—Steve Pouliot flipped over 360 Sam Rivers (Multiplex hologram). Selwyn comes off as the great genius—best friend of Salvador Dalí. He takes credit for what he and SOH (School of Holography) did and expects Hollywood not to find out. Ha! "Do you have any competition?" "Ah, no." If they only knew—Chris Outwater, T.J. Jung, etc. Selwyn says he introduced holography to Dalí when Cecile told me it happened through someone else.

I called Richard Arons and shared my frustration about Richard Aaron. He told me that he doesn't even think Richard Aaron knows the Ohio Players. "Go to the D.A., that's your best move."

Monday, June 17, 1974

Meeting at 1526 N. Fairfax—Ron Watkins, Mike Aaron (Aaron Brothers stores), Selwyn, Phillip—discuss Lumilar manufacturing. Ron took two of Dennis Pelletier's prototypes to cost and figure engineering. Selwyn and Phillip jumped on me cause I had Ron sign for prototypes. I handed the paper to Selwyn who handed it to Phillip, who gave it to Ron. Very heavy! I fell apart. Worried about Richard and my money. Extreme anxiety. Selwyn taking his frustration over Cecile out on me.

As I go through my diary I recall 99 percent of the people I've mentioned. Phillip, however, eludes me. I can't place him. All I know is that he is very much a part of this chapter. He obviously knew Richard Aaron because it was Phillip who called me at 6 am to say that Richard told him to tell me to meet him at Bobby Roberts' office at 1:30 so he could hand me five hundred dollars.

I had to believe that Richard would be there with cash. Too many people in the music business knew he'd written a bad check. I waited for an hour and when it became clear that Richard wasn't coming, I was about to walk out when an empathetic receptionist told me I had a phone call.

"I don't have it. I couldn't ask for an advance. I didn't want to look hungry. I'll be there tomorrow," Richard pleaded.

Tomorrow and tomorrow and tomorrow. Nothing had changed. No money was forthcoming from Richard Aaron.

To make matters even worse, when I arrived at my father's office that Friday, I discovered that my Jupiter 5 bank account was overdrawn.

For once my father was there for me. He gave me a check for $300 and said, "It would be cheaper for me to retire you."

I rushed to the bank in Brentwood to find out why I was overdrawn on my business account. "You returned a check to the Diffraction Company for $179. That's just not possible," I thought out loud, my voice registering hysteria. "I know what's in the Jupiter 5 account. Down to the penny."

The grey-haired, fifty-five-year-old department manager, was reluctant to go over my business account. She smiled and sighed and smiled and sighed and finally got into it. Let's face it, I was already very much on their radar.

"How'd you get that figure?" she wanted to know.

"I wrote it, but they weren't to cash it 'til I told them. It's taken out later, see?!" Her eyebrows raised and lowered as she worked her adding machine. Next, mister straight-back-crew-cut bank manager arrived to help sort out the error. It looked like it was the bank's fault and the manager started sweating.

Ah-ha! Someone at the bank had written the wrong figure in one column. I winced and laughed. I didn't make a mistake, they did. Once that was out of the way I showed them a printed hologram and asked them who at the bank to contact concerning holograms for the Bicentennial.

Monday, June 24, 1974

Called Richard and he said, "I'll be at your bank before twelve."

"Do you have the money? I'll meet you."

"Well, I'm meeting with Barbra Streisand's manager at the CMA building.

"I told him how to get to my bank from there. "Do you have the money?"

"I have a check for $900. I'll meet you there and give you the difference."

Not unexpectedly, Richard never appeared at the bank. By Tuesday I had to accept the fact that I was never going to see my money so I went to the Hollywood police department and filed a fraud report. From there I went to Jeff Allen's to meet Michael Foster, who told me about large sheets of diffraction gratings that he was making. I saw a rainbow diffraction grating pattern on rubber vinyl. Foster assured me that he could produce this beautiful effect on any number of surfaces.

Convinced once again that the record industry was going to embrace this visual pop of spectral magic, I left with samples and a price list. I didn't like or trust Jeff Allen and I was about to validate those feelings.

I had spoken to Chris Outwater about Michael Foster's printed holograms. Chris told his book publisher, who got in touch with Jeff Allen and ordered holograms as a promotional giveaway. Hindsight being twenty-twenty, I should have taken Michael Foster's printed samples to Chris' publisher and taken an order.

My lesson was clear: if you tell me about someone who can do something, I no longer need you. Respect for one's fellow adventurers was just not part of the landscape. A few years later Jeff Allen learned that lesson himself when Michael Foster no longer needed him. Dennis Pelletier had been working on designs for the VW bug and Mike Van Horn had come through with his car. I brought the Lumilar and, elated, we spent every waking moment cutting out stars and Liberty bells and a formidable eagle for the hood. Sometimes TONTO, Malcolm's synthesizer, would start playing all by itself. It felt both spooky and exhilarating, as if we were being encouraged from the great beyond.

> Wednesday, June 26, 1974
>
> Began work on the Lumilar car. Dennis Pelletier washed it with acetone. We scrubbed the running board—the whole car. We are having so much fun with the VW Bug and what it's going to look like.

Disney's car caper comedy, *Herbie Rides Again*, was riding on the popularity of the *Love Bug*, released in 1968. The studio needed the sequel to be successful.

Thursday, June 27, 1974

Richard called to say that he was on his way to my bank to deposit $100 into my account. He told Ulrich Jones (bank manager) that he was coming today with $150. I called Jones two and a half hours later—4 pm and no Richard and no money.

Friday, June 28, 1974

Call Mike Van Horn for letter stating that it's alright for us to use his VW in the Disney contest. Selwyn bollocksed things up with Burton Holmes Travel—Bob Hollingsworth. He said he made a Multiplex hologram when he can only make one with the SOH (School of Holography) equipment. Bob Christiansen like a bum, doing things here and there, not having money and expecting others to support him. I went to Malibu. Dennis designing and me cutting. Deborah Lotus arrived with pizza. I adore Dennis. We share a fly sense of humor. He disappeared for two years living in NY in a perfect spot with Osley pure in gin and taking one more drop each day till he reached 47 and walked in the woods and spoke to deer. He's been part of the Andy Warhol scene, all high art scenes.

What could be more Disneyland-American than a Bicentennial tribute? It was perfect. We worked day and night, night and day, Dennis designing and the rest of us assisting while videographer Paul Holman documented our progress for posterity.

We were convinced that we were creating a new art form called Lumilart. After all, Malcolm and his partner, Bob Margouleff, had advanced contemporary music with TONTO, so why couldn't we add a new dimension to art? We were on the cutting edge of a new medium, and Paula Cecil started crafting Lumilart art pieces.

We even decided to highlight the Bicentennial theme by wearing blue jeans, cowboy boots, red, white and blue scarves, and vintage political buttons. We were ready to win a new car.

Sunday, June 30, 1974—Herbie Day at Disneyland

Awoke at 6:30 am. Woke everyone. Malcolm slept on. Dennis, Jeffrey, Rod, Milton (Cecil) and I found our way to (Anaheim) 1000 tacky, comedic, nostalgic and patriotic VW lines. So sure we'd win. I was laid back. We finished the car and judging began.

Our patriotic little VW bug stood out from all the other bugs. At first the day was overcast and we needed bright sun to really dazzle. As we waited and waited and waited for the judges to come around, we each took a turn checking out our dull competitors. Finally, the sun broke through and suddenly we looked positively radiant against the sizzling black asphalt. By comparison, the other cars looked like cute homemade attempts—eyelashes above and below the headlights, or Herbie's number with flowers. My team exuded confidence.

We patiently shifted our weight from one foot to the other as the sun cast reflective rainbows. "Oh, that's what Akron puts on their toilet seats," yelled a competitor. "Yeah, you're right!" came the reply. Was this a judge? Who was referring to our expensive Lumilar as something a discount store like Akron put on their toilet seats? We stood out, alright, but apparently not in the Disneyland Orange County approved way.

The winner was Nostalgia, an homage to Disney characters achieved using chicken wire and paper. We didn't even win a prize in our own category and we were thoroughly exhausted. We drove the fabulous little bug back to Malibu and crashed hard.

CHAPTER 19

UP IN SMOKE

Monday, July 1, 1974

> Sleeping at Malcolm Cecil's at Point Dume. 7:45 AM—the phone rings on and on and on. I finally answer and it's my mother hysterically asking for me. She puts daddy on. "I hope you have a good reason for calling at this hour." "I do, the office burned last night."

To this day the thought of this event makes me quake.

I calmly put my things together and drove to town…expecting the worst. The fire was set on the second floor—my father's personal office that he was letting me share. The arsonists called the fire department to report a fire so that the whole building wouldn't burn. It was localized (upstairs) and the only real damage, aside from the structure, was to my personal belongings and to all of HCCA's McDonnell Douglas holograms.

My diary papers were alright—strewn around. Closet broken into. $7,500 worth of Sun Ra's destroyed. Teapot hologram that Lloyd and Peter Van Riper shot is broken. I don't know if they're after me or daddy. The interior of the closet was destroyed, taking my antique lace, velvet ribbon, books, Channing's (a large Channing Peake painting) taken out of frame—find on the floor. Frame melted. Air conditioner melted.

Carpet gone. Couch, all furniture destroyed. Tonight I discovered that checks were pulled out. It was me they were after. Outer office, closet, desk, drapes, carpet burned along with the gilt frame Richard Harris gave me, and a photo from Acapulco. Miracle my writing was saved.

The fire devastated me as much as it burned not only tangible items, but also my relationship with my father.

As I've pointed out previously, my father and I had a complicated relationship. On occasion he was supportive. For example, when he gave me a check for $300 to cover a Jupiter 5 overdraft. He understood how upset I was and with the swipe of a pen he fixed it.

Only this was different. His construction office had been in the same place since the late 1940s. The building took up half a block with a deep backyard for construction equipment, trucks and a gas pump. It was on an industrial street, which was surrounded by a neat, lower-middle-class, black neighborhood. If someone wanted to steal something, all they had to do was break in and race outside with office equipment. It didn't take long for me to realize that the fire had nothing to do with my father or the neighborhood. It was specifically directed at me.

After carefully analyzing everything that was destroyed, as well as the elements of the break-in, and making a phone call to the fire department, I came to believe that it was Richard Aaron's way of getting me my money. I believe that he thought my father had plenty of insurance and that he'd replace what I'd lost; only that didn't happen.

The fire had been a red flag to a bull. My father saw me as the villain, and if he hadn't allowed me to share his office, this would never have happened.

True as that might have been, he used the insurance money to rebuild the top floor, making it larger, and more modern and luxurious. Zero insurance money was my punishment. Oh, I shared my suspicions with the police and the fire department, but neither was able to find the arsonist or link anything to Richard Aaron. To me it felt like they just didn't care.

The repercussions reverberated from one coast to the other. HCCA didn't have the McDonnell Douglas holograms insured so they were a total loss. Instead of getting sympathy or empathy, I was blamed for their destruction. HCCA also was counting on my father sharing his insurance money.

Not knowing what to do or which way to go, I turned to William, my astrologer, for guidance. He looked at my chart and told me that three men started the fire, a Leo, a Capricorn and a Virgo. He kept asking me why I'd waited so long to have a reading. Apparently, all the bad signs were poised in my chart and for the first time, I sensed that he was shaken by what he saw. "You have to incorporate between July 24th and 28th or…" his voice trailed off.

Today if you experienced this kind of a tragedy, you'd go on Kickstarter or start a GoFundMe page. In 1974 they didn't exist, and a bank would definitely not give me a loan. William's warning motivated me to pick up the pieces and move forward at warp speed.

Tuesday, July 2, 1974

1 pm Edith Wiley—Egg & Eye. Edith cordially welcomes Lumilart as a new art form. Egg and Eye has no money at moment and would I agree to take consignment? Yes, and she took $63 worth of Lumilar jewelry. Not up for dichromates "Yet." I agreed we'll be in the Xmas catalog. Arson investigator says door to closet open so obviously they opened before burning—knew about holograms.

On the bright side, I occasionally got to drive the Lumilar VW and it was a hit everywhere I went. That and holograms buoyed my spirits even though I knew both Richard Aaron and my $750 were gone forever.

Saturday, July 6, 1974

Rich Rallison—he showed me how to make a printed hologram master (like Lumilar) acetone on vinyl or acetate: pressed into acetate as it goes thru a washer ringer. So simple! Easy to see how Mike Foster turns out "secret" printed variety. Rich says Mike is the only one who knows how to make "masters." I feel so

> strange, so removed living at various people's places and owing so much money. Knowing "all the right people" but waiting for a deal to jell. All so much mystery and not much time to judge the good guys from the bad.

At this point I'd fully embraced the persona of Laser Lady. The name felt right, and I loved hearing it. Wearing bands of Lumilar bracelets cascading from wrist to elbow, clothing airbrushed for me by the Laguna Rainbow artists, and driving my big white Toronado with lightning bolts on the sides helped cement my new identity. When I was in Laser Lady mode I felt confident, even invincible.

I hadn't given up screenwriting, but with everything that was going on, I had to make enough money to pay living expenses and vendors. Development deals and options could take weeks or months to negotiate, and I needed immediate revenue.

Rich Rallison made a diffraction grating hologram on clear vinyl, then on clear and black mylar. He explained that it could be used as gift wrap. In fact, that's what you see today when you buy shiny (silver or gold) rolls of holographic Christmas wrap. Raising money to manufacture it by the yard became our next topic of discussion. I was surrounded by successful people in the music business who I hoped would be interested in investing. Every time a holographer showed me something new, I was ecstatic and had to share it with a kindred spirit.

I'd become smitten with Dennis Pelletier, and although he professed to be bisexual, I chose to believe that I could win him over. I'd dubbed him Amir Façade, my brilliant Lumilar artist. He in turn wined and dined me. Insightful friends tried to warm me, but I chose to be a romantic fool with him as my anchor. And, boy, did I need an anchor.

The winds of change kept sweeping through, eliminating some players and introducing others. Shangri L.A. Ranch, aka the Pyramid house in Malibu, was finished. Zar and Silverbird had taken a gypsy wagon to parts unknown, somewhere in Washington or Oregon. Larry Spann closed his West Hollywood hologram shop, and Bob Christiansen disappeared. The holographers from Utah were now front and center at Burton Holmes International.

Donyale Luna had a new project, a TV special with her climbing the Empire State Building. In Luna's singularly tall, painfully thin, willowy way, she was trying to make the world see its beauty through her. However naïve this may sound, Luna was sincere. She saw "Fly or Die" as her mission and message.

> Saturday, July 13, 1974
>
> Hippo video—8:30. "In" party for the new Hollywood. Dennis (Pelletier) and Poli (Cecil) impressed by New York vibe. Candy Clark called so I asked her to join us. It took eons to get there because Dennis Lumilared his green caftan for me. Great fun to see so many old friends. Goldie Glitters (transvestite from Paradise Ballroom), Burton (Gershfield), Barry Gott looking so weak it scared me, but his work (sculpture) is successful—Candy's going down to look at it. Robert Patrick—New York eccentric—Corey type playwright. Presented his plays at L.A. Mama, old friend of Dennis'. Viva, who didn't remember Dennis (from Warhol studio) when prodded, said, "No, I didn't remember you. I was horribly depressed in those days and I've almost totally blocked them out." (from Chelsea Hotel in NY). Amazing to feel successful. For once I received credit in the program though I did little—one might thank cumulative effort. The party was so reminiscent of growing up with the Coreys.

Word came back from my lawyer that it would cost $400 to trademark Lumilar and $800 to incorporate Jupiter 5. There was no way that either of those critically important things could happen in July. And, to frost the cake, I received a shipping notice from Greyhound to pick up twenty-nine of Cecile's boxes and rolls of Lumilar in downtown Los Angeles for $263.65.

Wednesday, July 17, 1974

Sharks on one side, piranha on the other & Utopia in the distance.

Along with Andy Pfeffer at Mitchell, Silberberg, and Knupp, I now had a second lawyer named Ron Suppa. Ron was young and enthusiastic, only this call was anything but what I wanted to hear. "Without a purchase order Playboy won't pay for anything."

Ten minutes later the phone rang and it was Newsweek wanting to interview me via Beth Ann Crier (*L.A. Times*). I was shocked on both counts.

I was on an emotional roller coaster. The ride went up to the sky and then plummeted to the bottom of the Grand Canyon. Every once in a while things would go smoothly and I'd be lulled into a false sense of security.

The School of Holography had opened a branch in Santa Monica, and as usual they were struggling. Gary Adams, an original member of Lloyd's San Francisco crew, had relocated to set it up and teach. Ironically, people were wowed by 3-D images, but they didn't necessarily want to pay to learn how to make one.

I'd managed to rent an apartment in Santa Monica near the pier, within walking distance of the school. It wasn't my first choice, but I was running out of options, and Cecile needed a place to crash while the two of us tried to save Jupiter 5.

I could see that Gary Adams, although smiling, was looking especially gaunt. The Santa Monica school's phone was about to be turned off, signaling a replay of San Francisco. I bought a small, old, classic hologram of a butterfly, poet sparkle plates, and a spaceman. He needed money and I desperately needed holograms to sell.

Understanding how shaken I was from the fire and its consequences, JoEllen Lynott flew down from Seattle to cheer me up. Her timing was good because my circle had expanded to include a charming Ghanaian-born English percussionist named Rocky Dzidzornu, who I'd met at Malcolm Cecil's.

What Rocky lacked in height he made up for in magnetism. He

was charming and poetic with a wicked sense of humor. When he told me his father was a witch doctor in Ghana, I had no trouble believing him. He was a superstar drummer, so when he asked Dennis to turn his drums into works of art, we were off and running.

Hanging out with Rocky led to hearing bands, meeting musicians, and turning them on to Lumilar and holograms. JoEllen and I were treated like rock stars, driving around in my friend's limousine, being ushered into private dining rooms and backstage at concerts. It was the perfect way to reciprocate after she and her husband had so generously invited me to Hawaii.

I was excited to introduce JoEllen to Julia Chuhrayla, the singer who Richard Arons, co-manager of the Jackson 5, was now touting. Her unique sound, her throaty-Joplin voice and songwriting abilities so captivated him that he bought her a very expensive Martin guitar. We all thought Julia was about to sign with a major label.

One night, driving back from dropping Julia off at Richard Arons' sprawling neo-Roman house in Trousdale, high in the hills bordering Beverly Hills, my brakes started smoking. The streets were so steep they were practically vertical, and I had all I could do to put the heavy car into low and squeeze up against a curb. The brakes were gone but miraculously we stopped without injuring anyone or hitting another car. Triple A towed us to a gas station on the Sunset Strip.

"What's that stuff?!?" the station attendant barked, pointing an accusing finger at my lightning bolts. It seems someone had hired him to make Lumilar key chains and then wouldn't pay him. More accurately, couldn't pay him. I made one call to my friend Burton and within minutes he whisked us away to Studio One in his limousine. Somehow there was always a silver lining.

Monday, July 22, 1974

Fred Slatten flipped over gold (Lumilar) shoes and said Playboy had been into film and he'd given them to them to shoot. "You wouldn't be mad if I paid you, would you?" A smile flashed through my entire body as he handed me a check for $50 and Elton John's boots.

Owing to the full page in the Sunday *Los Angeles Times Magazine*, Lumilar shoes had stepped into the spotlight. Jewelry was still popular, but now I was taking orders for custom jobs. In some strange way it was lucky the *Times* editor had shifted from jewelry to shoes, otherwise I would never have met Fred Slatten.

Dennis finished Rocky's drums in time for Minnie Riperton's opening at the Troubadour, and *Newsweek* was there to capture it all on film. The drums lit up the stage along with Riperton's otherworldly, soulful voice. Nobody could hit a high note like Minnie Riperton.

Thursday, July 25, 1974

Running, running, burning out—no real office—harassed by father and given continuous advice from mother. I'm going nuts—feel like a rat in a maze searching for path that leads to quiet and sanity. Always going opposite direction into uncontrolled situations. I went by Fred Slatten's to pick up more shoes and Fred wanted to discuss the high cost of Amir Façade's custom work. "You're charging too much, Linda. You need to bring the price down." Hugh Hefner was interested in ordering multiple pairs, and he wanted a better price. "Hefner has the last word," Slatten warned.

Monday, July 29, 1974

The Senate voted second article of impeachment for Nixon.

The times, they were turbulent, and this had certainly been the most turbulent month of my life. The fire had destroyed not only business and personal items; it had incinerated my relationship with my father. I'd finally taken an apartment in Santa Monica and Cecile had piled in and taken it over. We had no money to trademark Lumilar let alone incorporate. Partying was great if you had something to celebrate, and someone could always think of something. JoEllen had a wonderful time, even going to a Minnie Riperton recording session.

Once she headed back to Seattle, I took a deep breath, and began seriously assessing the damage. It was clear that there was no love lost between HCCA, Cecile, Selwyn, and myself. A number of full rolls of Lumilar were still in New York, and Jody Burns was selling it by the foot. Thanks to Playboy refusing to pay the invoices, I needed those rolls to generate cash flow. Jupiter 5—meaning me, since Cecile had no money— owed the Diffraction Company thousands of dollars. I had to come up with a plan and I had to do it quickly.

CHAPTER 20

WHAT'S MINE IS *NOT* YOURS!

Wednesday, July 31, 1974

Cecile agreed—offered to take 25 percent instead of 50 percent (of Jupiter 5)—Dennis (Pelletier) can have 25 percent if he fulfills his art and treasurer duties. Barry's (Gott) studio over theatre—5000 sq. feet of artwork—undulating pictures and kinetic sculptures. Thai weed. Candy (Clark) leaning back innocently listening to Barry's sophisticated rap on noetics, electronics and life. She looked at all his pieces, made her selection and said, "You know, this is my first piece of art." There I was between this genius Barry Gott and this lovely actress, Candy Clark, being the catalyst in a youthful, significant art transaction. He called Candy yesterday, but got her service. Barry's phone is now turned off. I called Candy and arranged the trek to Hermosa (Beach). This was progress since Candy had been spending every day with Kim (Milford) who's always "too tired" (to drive down to Hermosa Beach).

At seventeen, Kim Milford joined the original cast of *Hair*, and went on to play Jesus and Judas in *Jesus Christ Superstar*. I had originally met him at a party at Bob Margouleff's house in Malibu. I'm guessing that I met Candy Clark through Kim.

That winter Candy had been nominated for a Best Supporting Actress Oscar for playing Debbie in *American Graffiti*. She was interested in art and I was happy to introduce her to Barry Gott, an eccentric artist who was using multicolored geometric Plexiglass shapes and textures connected to tiny blinking LED lights.

These futuristic sculptures were housed in tall rectangular Plexiglass boxes with a clear top and an opaque bottom that concealed the electrical wiring.

They were complicated, cleverly engineered kinetic works of art, and guru Barry wore the hippy genius mantle well. His warehouse/studio was hoarder heavy, with electronics, multicolored pieces of plastic, scraps of wood, paint, glue, and miles of wire. When the artwork blinked and flashed in discordant patterns from one end of the cavernous space to the other, the effect was both mesmerizing and overwhelming. The massive clutter gave me claustrophobia.

Candy made her selection and we headed back up the coast. For me this was a momentous occasion. Selling an emerging artist's sculpture to a newly minted movie actress was heady stuff. I liked selling and I loved art. The combination was powerful. I could bring people together, and in the process, move a new art form forward.

This experience solidified my desire to become an art dealer. I had a passion for it, and at the end of the 1970s I formed Holographic Management Associates with Danna Ruscha (Ed's wife) and Randi Foster (Michael's wife) to produce the first fine arts gallery exhibition of holograms in Los Angeles: Holography '79 Series One, followed by Series Two at the Anhalt Barnes Gallery on La Cienega.

At this point though, I desperately needed a break, so when Dennis Pelletier asked me to join him in Hawaii, I jumped at the chance. He and some friends had rented a house on the Gillette estate for ten days. "Just get here, and everything else will be taken care of." How could I resist? He was charming and funny and talented—for me, a trifecta aphrodisiac. I managed to beg and borrow enough money for a round-trip plane ticket.

Saturday, August 3, 1974

Basket weaving at Beth Ann's (Krier). Up all night. Alan (Abrams) drove me to the airport. I made it to passenger check in—man looks at me and says, "How would you like to fly First Class today?" What a blessing! Met a writer—reminds me of Alice (Corey). She and husband fish at Catalina. "I bet your parents are proud of you!" Someday. Landed—Dennis and Terry brought me beautiful leis.

My first trip to Hawaii with the Lynotts was a perfect hotel holiday. This time, I was staying in a guesthouse on the Gillette estate in Kailua. That night Norman, the head gardener, and his cousin, Robert, dug a deep pit, which they filled with wood and black lava rock. A one-hundred-and-ten-pound pig was wrapped and buried along with Hawaiian sweet potatoes wrapped in banana leaves.

While the pig was roasting they made chicken with long-rice soup, salmon sashimi, and fruit salad in the kitchen. It turned out to be a delightful mai tai-fueled backyard luau for seventy-five guests.

Dennis and I played out our complicated relationship. Just as Paula Cecil, Malcolm Cecil's wife, had once warned, once we drew close he started to pull back. And, of course, my mother's threatening send-off echoed in my head. "If you go to Hawaii, something horrible will happen. I can tell. I can feel it. It's the worst thing you could do!"

In spite of negativity I persevered. I heeded Paula's warning and gave Dennis plenty of space. I made the rounds of local shops selling pieces of Lumilar, holographic jewelry, and packages of diffraction grating die cuts. I was productive, and being away from toxic relationships in Los Angeles had helped me swim back up to the surface and breathe again.

One afternoon a group of us went to lunch at a restaurant in Waikiki called Toulouse Lautrec. I met a young artist named Chang Young Cho and was so impressed with his work that I asked if he would consider trading holograms and liquid crystal jewelry for one of his paintings. He agreed.

Chang was twenty-six or twenty-seven, a Korean-born artist who was married with a child and insecure about his future. He felt that Honolulu was too expensive, as was San Francisco. He thought Oakland would be a better place for him to take his family.

> Thursday, August 8, 1974
>
> "Charlie," as he is called, is a Korean wonder. His paintings are simple, whimsical and still an organic compilation of our lives: what we are, what is around us and what is inside of us. Dennis and I went to Liberty House with jewelry. Assistant jewelry buyer flipped, says buyer will call me in L.A.

Dennis and I missed our 6:25 am flight to Maui but caught the next one. We joined his friend Buzz at the Pioneer Hotel, which was more bed and breakfast than hotel. For $11.44 per night we had three twin beds with a bathroom down the hall. Buzz finally struck out on his own to search for his girlfriend, who was somewhere in the enclave called Lahaina. Dennis took a nap while I went from shop to shop selling my magical holographic wares.

> Saturday, August 10, 1974
>
> Dennis and I finally got to be alone. Went to the Lahaina Broiler as the sun set over the island opposite. It was the perfect setting, the perfect feeling and delicious sashimi, garlic toast, mahi. Went back to the bedroom without a bathroom and Cat Stevens was blasting from the room on the left—got stoned and discovered a much deeper connection. Could hear piano player's tunes rising from the bar beneath us. Switched mattress to make a double bed.

Sunday Buzz, Dennis, and I drove to Hana, a fifty-two-mile, half-day trip winding through lush green vegetation splashed by waterfalls. It rained on and off on our way to the Seven Sacred Pools, a miracle of nature with crashing clear aquamarine waves, verdant hills with palm trees to the right, and lush tropical mountainous forests to the left. We finally headed towards Lahaina.

Prior to leaving L.A., Eddie Auswachs had prompted me to get in touch with a friend of his, an actor, singer, producer and former star of the musical *Hair* named Red Shepard. He'd given me his number, but I saw no point in calling to say, "You don't know me, but…"

We were driving towards Hana along a narrow winding dirt road when we found ourselves bumping along behind an open Jeep with gigantic truck tires. The driver had a long flying bush of bright red hair.

"Do you think it's Red Shepard?"

"Ask him!!!" We honked and I screamed, "Are you Red Shepard?"

The Jeep stopped. "Yes," he yelled, and before we knew it we were sitting on his veranda sipping hot tea and munching on Maui Wowie mushrooms from cow pies collected from his backyard.

Red was wonderful and a little lost. Hana was a beautiful, completely off-the-beaten path place to live. Red wanted to work, but he couldn't quite find a project worth pulling him back to civilization.

"I'm the one who turned Eddie onto the idea of using projected holograms for his play." Red looked at me. "What can I do for you?"

I was stumped. I had no idea. Inviting us to his comfortable home and pointing out rainbows forming in the mist was more than I could have wished for.

"I saw *Hair* on Broadway in 1968 and it changed my life! It liberated me! I was a theater major…" I rambled on because for me this was a special moment. For Red it was just another conversation cementing him in limbo.

On Monday I sold more bracelets, stick-ons, and yardage to a shop in Lahaina. It felt like an affirmation. The air was intoxicating and Dennis and I were in harmony. How long that feeling would last was going to depend on what happened when we got back to Los Angeles. Paula was furious with me and we both knew it. To allay my fear, Dennis asked where I'd like to have dinner the following Saturday night. To me that meant that he saw us as a couple. That our relationship was solid.

Dennis, Buzz, and I flew back to Honolulu. Before heading back to the mainland, I took Dennis to meet Dr. Lugaivi and experience his powerful light show laser at the University of Hawaii.

> Tuesday, August 13, 1974
>
> Went to University of Hawaii to meet Dr. Lugaivi and his asst. to see helium neon turntable die laser. Ron Goldstein and Ed (Auswachs) doing New Year's show at Bishop Museum. Dennis overwhelmed by laser's beauty.

From the University we raced to the airport. My flight was at one and I arrived on the dot, but as luck would have it the plane was late and I boarded with mixed feelings. In my gut I knew that returning to Los Angeles would test my relationship with Dennis. After all, he lived at the Cecils.

Then, a couple days after touching down, the phone rang, and much to my surprise it was a chirpy Paula Cecil telling me she needed Lumilar for an art project and would I come out to Point Dume. I rationalized her invitation as a truce, and when I walked in, I was greeted by Bob Margouleff and Malcolm, who seemed genuinely glad to see me. In retrospect maybe they just couldn't wait for the main event.

Paula took me aside and showed me her "Lumilart." "How many times did you sleep with Dennis?"

I said nothing.

"Well, that's it then! You did a lot, I can tell! What do you have that I don't? What?!" She raved on and on, her full, English voice resonating with fury.

"Paula, your work is great!"

"You really think so?"

Paula knew her place was with Malcolm, but being married to a workaholic had opened the door for her attachment to Dennis. Malcolm, the musical genius, wasn't threatened by his wife's relationship; in fact, he undoubtedly welcomed someone who could be an assistant to him and a companion to her. What he hadn't counted on was the intensity and tenacity of her attachment.

"You're never to spend the night in this house! Do you hear me?!"

I should have lied. I should have said that Dennis and I had never slept together and she would've been happy and I wouldn't have created an adversary. Lesson learned. From then on Dennis and I would have to be strictly business: Amir Façade creating custom designs for Jupiter 5.

> Friday, August 16, 1974, 11 AM
>
> Ron Suppa—discussed incorporating and trademark—gave him $300 to start trademark. Lunch at Swiss Café with Mayo Simon to discuss sequel to Westworld. He knows little about the future. Cecile looked the part in a black jumpsuit, silver cap and 1" wrap around sunglasses. "How much would you charge for an hour or two of consultation?" Cecile and I vibed each other across the table.
>
> "Two hundred dollars."
>
> "I'll tell the producer." Mayo was won over by coffee. Overwhelmed by auras, biofeedback, Reality! Went to office.
>
> Richard Arons called, "You can come get the shoes."
>
> "But I brought them yesterday."
>
> "The boys shoes!" Then he cut me off while he put me on hold. Got him back, zipped to his office, picked up six pairs of (Jackson 5) shoes. For the first time my timing was perfect.

Richard Arons had come through with the Jackson 5 girls' tap shoes, and after seeing how cool they looked, the boys wanted theirs done too. They would be wearing them on stage at the MGM Grand in Las Vegas and that was huge. Michael Jackson was sixteen, and he and his brothers were touring the world with their perfectly synchronized moves and chart-topping vocals. Dancing Machine was a number one R&B hit, and number two on the pop singles chart. Between Elton John's boots, the Jackson 5's tap shoes, and the Bicentennial VW bug, Jupiter 5 was on the cutting edge of fashion and art, and I needed Dennis to put the post-Hawaii soap opera behind him. I was still willing to offer him 25 percent of Jupiter 5 in exchange for creative input and custom work, only our last exchange had been cool, if not downright chilly.

"You had a long face today," he'd observed.

"I did, and you're the reason why!"

"I am?"

"Before we left Hawaii you asked where I'd like to have dinner Saturday night. Now...I don't even know if we're still friends."

The blush was off the rose. In a bitchy tone he said, "How much do I get for the shoes?!" He knew exactly how much I was charging clients and how much he'd receive. In the beginning it had all been fun and games. Now that it was shape up or move on, there was nothing to think about. Of course, Dennis was going to take his job and home security over someone surviving by her wits. He and Paula had a strong bond. What was I thinking?! I think the answer is: I wasn't thinking.

Sunday I called Paula "As soon as the Jackson 5 shoes are finished, that's it—you don't have to worry, Dennis is yours. You win."

The turnaround was fast, almost too fast. The boys' shoes were ready, and when I drove out to Point Dume to pick them up, Paula was nice, but cool, and Dennis was reserved. All of his warmth was directed at her. He hated the white patent leather tap shoes and only cut the accents and sealed them with Varathane, instead of decorating the whole surface. What made the Jackson 5's dancing feet spark kinetic splashes of color depended upon the amount of diffraction grating material on their shoes. By only using a small amount, the effect would be minimal, and the Jackson 5 did not do "minimal."

As I drove away I felt a deep sense of loss. Yes, Paula had won. Dennis seemed happy to be home and to be with her. We would all continue to be "friends" because that's what sophisticated people did. They sucked up disappointment and moved forward.

CHAPTER 21

STEPPING OVER LANDMINES

Growing up an only child of a mother possessed by narcissistic personality disorder was always a challenge. She was always right. It didn't matter what it was. Where was I going? To whom was I talking? What was I wearing? She knew best, and if I defied her, I got a crooked witch's finger along with an ominous, premonitory warning.

Well, this time, I'm afraid she was right. Going to Hawaii had inflamed Paula Cecil, and in turn, Dennis Pelletier and all of his cronies. I was the villain because I'd had a good time, and Paula was determined to cut me out of Dennis' life. I heard that she had begun looking for other sources of Lumilar. Fortunately, for me, there weren't any selling the same quality and pattern variety.

Dennis' days as the creative director of Jupiter 5 were over, and I had little interest in mastering the design and application of adhesive-backed holographic foil.

I found myself at a crossroads, immersed in all things holographic and otherworldly. My vivid imagination had been unleashed, and at my core I believed I could see the future. As time and original screenplays proved, I'd found my niche as a futurist. What I was writing as science fiction then is now science fact.

August 22, 1974

Rich Rallison—8 PM. Took Dennis (Pelletier) to meet Rich. Dennis cool, almost cold. Walked into Rich's. Waited. He (Rich) arrived with a broken leg. No cast but doing fine. Rich didn't like Dennis, wouldn't answer his questions without negative inflection. Dennis took an egg-shaped dichromate jar. Stopped for a drink. Neither of us had more than $2. Used all of it for Margaritas. Had pictures from Hawaii. Brought back beautiful memories less than a week fresh. Now, I feel sick, empty.

On Sunday Cecile Ruchin and I set up a table at the Venice street fair, and once again we waited for bright sunshine. It wasn't until the afternoon when the Lumilared VW bug arrived that people started getting excited about our rainbow emitting jewelry and magic wands. Crowds were magnetically drawn to the bug, so we moved our table closer to it. That didn't go over well with the officials, who told us to move back to our assigned space or leave! In the end we sold enough to justify our frustration and exhaustion.

There was no getting around it. Like it or not, Cecile and I were still partners. She had gone her way and I had gone mine. She was living at the Santa Monica apartment, and I was staying with a variety of friends. We were both responsible for paying the Diffraction Company back for the rolls of material Donyale Luna had convinced Jody Burns to order. A mind-blowing event was the only option.

My candle in the dark was actress Candy Clark, who liked what I was doing with art and holograms enough to offer to co-host a Jupiter 5 fund-raiser. She would engage the Hollywood community to buy tickets and turn out for a once-in-a-lifetime technological extravaganza, while I would line up holographers, artists, and musicians to exhibit and entertain.

Word went out that we needed a large venue, and a friend who lived at an upscale Point Dume trailer park offered us his spacious clubhouse overlooking the ocean. The rental fee: fifty dollars. Candy contacted the owner of Vendome Liquor, who agreed to donate a couple cases of champagne. We were on a roll.

Monday, August 26th, 1974

Went to Barry's (Gott) studio with Candy. She had an interview so we zapped down to Hermosa (Beach)—ran upstairs and knocked. A very stoned Barry slowly opened the door. His red-veined map-like eyes flashed towards communication as his thin body swayed. "The benefit's going to be September 28th, and I want to feature your work!" "Come on, Barry! Get it together!" And Candy and I were gone. I took some more magic wands. I don't think Barry will remember the experience as real, more like a fast dream.

Of course, as I've said before, holographers were loose cannons. They could tell you they were participating, then never show up. It was a game of blind faith. Candy was the charm. After *American Graffiti* she was both well-liked and respected around town. If anyone could motivate George Lucas and Steven Spielberg to drive out to Malibu to see holograms, she was the one.

It felt like the tide was finally shifting in my direction. Then Richard Arons called. His voice was agitated and resolute. "I'm not paying for the boys' shoes! All the holography crap peeled off!"

I knew that the only way the material would come off was if the Jackson 5 boys used force and peeled it off. It had a strong adhesive backing, plus, after it had been cut and adhered, it was painted with a protective coating of Varathane.

When I hung up I knew I was going to have to confront their co-manager face-to-face. He ordered the customization and he had to pay for it. Dennis and his helpers needed to be paid, and if relationships weren't already strained, non-payment for the Jackson 5 shoes would be the final nail in the coffin.

Wednesday, August 28, 1974

Having a wonderful haircut when Paula arrives to pick up her garage sale purchases. Jeffrey says, "You know, Paula, you're one of my favorite people!" "Oh, what?? Ha!" "No, you're so dynamic—a great character." She gathered her new Lumilar-trimmed fish tank, goldfish, screen. Dennis called and I answered the phone. "Paula?" "No, Linda. Small world." I smile, Paula watching, glares as I effuse, "The director says he's given the script to Step Two Productions. The chief read and liked—now it goes to MOW (movie of the week) executive."

This smacks of a cat fight. Paula and I were killing ourselves over a gay man. And it didn't end there. On my way back to Santa Monica I stopped at Bob Margouleff's house to see Creed about an upcoming party.

"Can I see you a minute?" he said, leading me into a bedroom. "Linda, there's not going to be a party. Bob doesn't want you here anymore. Paula's really gotten to him." I was crushed. One minute I felt victorious, boasting about a director's positive reaction to a script I'd written, and the next I felt like I'd been steamrolled, crumpled, and exiled to Siberia.

I was so upset, that I lost my car keys and had to call a friend to pick me up. Bob Margouleff looked at me casually, never making eye contact. When I'd spouted off in Paula's presence about an impending movie-of-the-week deal, she dug her claws in. I thought I'd won a battle, but clearly, I'd lost the whole damn war.

I loved Bob Margouleff's parties. He always had the most interesting people and they were all drawn to holograms. This was a major loss.

Friday, August 30, 1974

The onion didn't help. Tried to make myself cry for Richard Arons. Between Judy's (seeing buyer) and Babe Rainbow (seeing buyer)—go to phone booth. I have 50 cents, no more—get my sadness together and the line is busy.

Went to Jackson 5 office. Richard alone in the conference room. Immutable. "I won't pay money for something that came off! It fell off on stage. It peeled off!"

"The boys peeled it off, didn't they?!?"

"Yeah, when it came off, they peeled it. I won't pay and that's it. It was an experiment and it didn't work!"

"What am I supposed to do, Richard? My artists won't do anymore work. I'm going to have to declare bankruptcy."

"Is that my fault? Is your financial problem my fault??? There, I'll give you a personal check for $25, and I don't want to ever hear about the shoes again." I stood there totally humiliated, wanting to tear up the check and punch him in the nose. Instead I stood there near tears, but needing the $25 too much to scream what I was thinking.

"Debbie," he yelled. She came in—"Did the boys have the shoes on when you saw the show (in Las Vegas)?" She thinks. Richard finds out she saw them wearing the shoes last weekend.

"Okay, I'm going to tear up this check and make out a Jackson 5 check for $75. You can redo the shoes."

I move the car so I won't get a ticket—go up and all smiles—"Jermaine wants his guitar done."

My relationship with Dennis continued covertly. It wasn't easy, and it wasn't without drama. It was simply designed to leave Paula out of the loop. Technically, in Hollywood parlance: We were all still friends.

Thursday, September 5, 1974

Ruth Tuska from Wisconsin has left her magazine, Views & Reviews, and has run away to L.A. to join Cecile, who she knows will help—i.e. protect her from her irate husband who wants to

> kill her. Who's her friend in L.A.? Sam Peckinpah. Like walking into a soap opera. She tells me, "I'm not here. You've never seen me. I've run away."

It was a crazy time. Three of Dennis' young, gay acolytes were helping me move into the Jupiter 5 store. They had also helped Dennis with the Lumilar designs, so I had some hope of them being able to continue that part of the business. At least they were fun and kept me laughing.

> Monday, September 2, 1974
>
> Went to Julia's (singer Julia Chuhralya staying at Richard Arons' house) for a swim and sun. She called Memphis to wish a friend at Stax Records happy birthday—found out Stax received a bill from the Hyatt House for $841. Richard Aaron completely crazy. I knew something was funny, but not this funny. He didn't even go to Harvard.
>
> Dr. Shapiro calls—he's Jackson 5's dentist—he's in Palm Springs with Cary Grant. He works at night as do musicians. He told Julia he'd take her to the flower mart, then fix her teeth. Joe Jackson has more than one family. Perfect irony. Michael most sensitive, virginal. Afraid girls only after him cause he's "Michael Jackson." Started a diet. I'm starving.

Dennis and Paula fancied themselves innovators of a new art form. They were Lumilartists. They imagined their collages suspended from the walls of hip art galleries. The concept was plausible except for one key element: Jupiter 5 had an exclusive agreement with the Diffraction Co. If they wanted to buy the material they would have to go through me. Edmond Scientific might have a couple patterns, but they weren't an open source.

Dennis was caught in the middle. He'd told me that I was the only person he really enjoyed being around for any length of time. Now that sentiment was gone, and he was bitter. He'd redone the Jackson 5 shoes, and they needed to be delivered. The only solution was to have Cecile take charge, and she did.

Feeling that we were finally on the same page, I dropped by the Santa Monica apartment. Zar and Silverbird had just arrived, happy and satisfied from their great God-led gypsy wagon trip to Oregon, and Ruth Tuska was still alive and in hiding.

Here's a lucky piece for you, Linda." Silverbird handed me a sparkling deep purple amethyst crystal.

"Thanks, Silver, but right now I need a money miracle. I can't even afford to buy stamps for the invitations."

Ruth brightened. "How many do you need?"

"Fifty dollars worth."

"You're in luck," she said, "before I left Wisconsin I took stamps. I can give you forty dollars worth."

Big sigh. A miracle! Anger and disappointment weren't going to solve any of my problems. The only way to rise like a phoenix was to raise money and prove to the Hollywood community that something very special was happening at Jupiter 5.

Forget the naysayers—bring on the movers and shakers.

CHAPTER 22

NINETEEN DAYS AND COUNTING

Monday, September 9, 1974

May Co, Bullocks, Paul Holman, Cecile, 10 AM store. Went to Fred Slatten with Dennis. We go to Fred Slatten's. Dennis has a nice meeting with Fred—doesn't say anything about copyright—cools out. All runs smoothly and we get paid. Fred mentions us getting a line of shoes for department stores to Lumilar for him. Great potential.

Toss the dice and see where they fall. Leading up to the benefit, that was pretty much what was going on with all the disparate factions. I'd been spending time with Rocky Dzidzornu, the short Ghanaian-born conga drummer with the colossal presence. I'd met him at the Cecil's and we'd clicked on a cerebral front. Meeting eye- to-eye we often picked up the nuances circulating through each other's brains, and although he was a consummate professional, there was always an undercurrent of naughtiness and mischief. He told me he was onboard for the Jupiter 5 fundraiser.

My business partner, Cecile, was both connected to and disconnected from her HCCA partners in New York. She always appeared to be busy, but I wasn't sure what she was busy doing. Missing deadlines had become a regular part of her modus operandi. Cecile's friend Ruth was still in hiding, but that didn't stop her from using every tool in her wheelhouse to help us prepare for the big event. We had nineteen days to pull everything together, and we needed all the support we could muster.

My biggest blighted hope was the lack of encouragement from my parents. My father blamed me for the office fire and would show no empathy or compassion. When discussing the fundraiser he went so far as to say, "Don't invite any of my friends! I don't want you getting their sympathy money!"

So, no insurance money, no "let me help you" money. Just basically, don't bother me and you can still use Richard M. Lane Co. as your business address. My mother's approach was, "I told you not to go to Hawaii!" so there was no empathy or compassion coming from her either.

Then there were the bright spots: metaphorical rainbows-like designer Holly Harp ordering 10,000 Lumilar sequins, and *Newsweek* photographer Lester Sloan making an appointment to see the almost-open Jupiter 5 store.

I was living in the moment, desperately trying to be all things to all people, and desperately needing a place to decompress. Luckily, in the midst of the madness my songstress friend, Julia, invited me to stay with her at Richard Arons' sprawling neo-Grecian/neo-Roman house in Trousdale. I liked to refer to it as the marble mansion. Richard was in Brazil on tour with the Jackson 5 and Julia wanted company.

It was wonderful. At the end of the day I could drive up the Trousdale hills and lose myself in millions of dazzling lights below. The view was reassuring and made me feel hopeful. I needed to be clear if I was going to be organized, and being organized was a prerequisite when co-hosting with Candy Clark.

Candy was a go-getter, organized and proactive. There was no way that I could slack off. She expected me to accomplish what I had agreed to and that was enough to keep me in forward motion.

Ivan Dryer, the founder of the phenomenally popular Laserium music and light experience, and later named "the father of the laser light industry," took down his private collection of laser photographs to exhibit at the fundraiser. In exchange, I loaned him a hologram.

Chang Young Cho, the artist I'd met in Hawaii, agreed to loan me several paintings.

> Thursday, September 12, 1974
>
> I worked at the shop while Julia stuffed invitations—Cecile brought Barry Gott who brought a baggy with a dozen Thai sticks, and Candy's amazing sculpture. Rich Rallison gave me 10 holograms. "This place is amazing!" (shop). Ruth on hands and knees washing the floor. Cecile disappearing, reappearing.

I began the day by hand delivering a number of invitations, and gathering more addresses. Networking 101.

Lester Sloan, the photographer from *Newsweek*, was meeting me at Jupiter 5 at 3 pm, and if he liked what he saw he'd schedule a photo shoot. We had multi-patterned rolls of gold and silver Lumilar for sale by the foot, along with Richard Rallison's pretty dichromate holograms—magical jars that appeared to contain a treasure trove of coins and jewels. There was a counter with hologram jewelry, packages of stick-ons, and a pair of Fred Slatten's platform heels. There were a lot of interesting things clustered throughout the cold empty space. Elevating holograms from novelty to art form meant creating more of a gallery. We needed large holograms and art on the walls.

The Santa Monica School of Holography was ten minutes from the shop on Ocean Front Walk, and surely with *Newsweek* in mind, they would let us borrow a laser or comparable light source. Carl, who was in charge, told me he'd have to call Lloyd Cross in San Francisco, and I'd have to ask him.

I viewed this as an emergency. Lloyd Cross viewed it as an opportunity. "I want Pam and Puppy back," he said flatly, referring to his 1974 rainbow hologram of friend Pam Brazier with a puppy, a trolley car, and the Point Star hologram. "For historical reasons."

"Alright, Lloyd, but I don't have it, Cecile does, and she's at some fucking convention downtown."

Eventually a deal was struck. Lloyd told Carl what to dictate and had me type up an agreement. No one would be allowed to photograph a multiplex hologram without specifically crediting Lloyd Cross and the School of Holography. It was now 3 pm.

I rushed back to Jupiter 5 with a multiplex hologram, a light source, and holographer Gary Adams. The store had been transformed into an upscale gallery. Ruth's curtains against the teak latticework polished the rough visual edges. On an adjacent wall, a colorful Chang Young Cho lithograph had been hung.

Gary set up a 360-degree multiplex hologram and we waited. People came and went. Candy Clark arrived to pick up her Barry Gott sculpture, and finally Lester Sloan turned up accompanied by a lady friend. Apparently, Nixon was seriously ill, causing Lester to be held up at the magazine. The photographer was skeptical at first, but little by little, visual by visual, he warmed to holography, and after an hour and a half he offered to shoot a special layout to help us.

> Saturday, September 14, 1974
>
> Lester (Sloan) spent five hours shooting Candy (Clark) for Oui. She made a (Lumilar) belt and a sculpture between addressing invitations. Julia's party was a big hit! Too much food, (Richard Arons) house so cold, people felt isolated. Candy came by briefly. Creed and Julia are my favorite Hollywood couple. Spoke to Buzz who told me that Terry (Andrews) invited him over to show slides. "Where's Linda?" he had asked. Everyone looked at each other and went blank... "Oh, I don't know." Sure. My eyes are beginning to open. Wide.

The following day there was a full-page Lumilar-shoe spread in *Home* magazine. Lumilar shoes, designed by Amir Façade. Dennis didn't even call to say he'd seen it. No matter, there was no time to ruminate. I had to take holographic material to the Tomorrow Show.

The Tomorrow Show followed Johnny Carson on NBC. It wasn't a witty, laugh fest like Carson; instead it was a one-on-one with interviewer Tom Snyder, whose guests included people like novelist Ayn Rand and John Lennon. Placing holographic material on a set or as a talking point would give Jupiter 5 exposure and credibility.

By now Lester Sloan was convinced that he was coming in contact with a group of enlightened futurists. I found him at Cecile's discussing auric drawings with Ruth, and how technology was going to save the world with Cecile. I joined them and he sat sandwiched between three raving ladies. Overwhelming as it might have seemed, by the time he left he was a believer. "I'll cover your benefit for *Newsweek,*" he promised.

We could tell that there was a long, hot story percolating in his journalistic mind.

"You're the most interesting people I've met in a long time," he told us on his way out. Mission accomplished. He was blown away, and so were we.

Candy got record producer Lou Adler to donate twenty-five *Rocky Horror Picture Show* albums plus five sets of tickets to the show. I received the first positive RSVP from my agent, Crayton Smith. Little by little everything felt like it was coming together.

I met with someone at Coherent Radiation to borrow a laser for the benefit, then I picked up shoes from Fred Slatten and had Edward, one of Dennis' acolytes, transform them. He did an excellent job proving that I had put way too much faith in Dennis' singular ability.

That night my lawyer, Ron Suppa, Candy, and I brainstormed over dinner. Candy's impressive list of donations was growing daily: ten passes to Universal, dinner for two at Dan Tana's, and possibly a round trip ticket to Hawaii. Candy was amazing. She had the positive energy and commitment of a dozen people.

Lloyd Cross agreed to loan us one 360-degree and two 120-degree multiplex holograms.

Saturday, September 21, 1974

Cecile piled large rolls of Lumilar into her borrowed car and dropped them off at the point (Point Dume). When confronted by my negative feelings she said she knows what she's doing, and she doesn't want or need me telling her what to do! Julia read my cards. Power struggle with Cecile. I trump, but it's a long road to hoe. There's a Lumilared Liberace billboard. Everyone thinks we did it. So many questions in the air.

With one week to pull everything together, Paula Cecil loaned us two of her large Lumilar artworks. This may seem counterintuitive in light of her animosity towards me. However, it makes perfect Hollywood sense. We were expecting a number of A-listers: actors, writers, directors, and producers known to collect contemporary art. If Lumilart was introduced at the trendy fundraiser, people might buy a piece, and there might even be a gallery connection.

Candy reported that she had the Starks—Wendy and possibly her parents, Fran and Ray Stark—Julie Christy, Sally Kellerman, and Superman's future girlfriend, Margot Kidder.

I was driving all over town gathering invitations and checking in with people when I stopped for a red light at the busy intersection of Santa Monica Blvd. and Sepulveda. While waiting for the light to turn green, I heard brakes screeching, followed by the volcanic jolt of an Econoline van crashing into me. The van sent my car into the car in front of me, and like a house of cards, we had a five-car pileup. My car was totaled, and I had the mother of all whiplashes. As the day wore on, my balance was so out of whack that I accidentally fell into a mud puddle outside of the Writers Guild.

We were now in the home stretch, and like the Energizer Bunny I had to keep going. Candy had me stop at producer Si Litvinoff's house in Malibu Colony for his guest list. What could be better than inviting people who lived down the road from the event. They could be charitable, drink champagne, and make it home in ten minutes.

Sunday, September 22, 1974

Party at Sammy Davis, Jr's. Altovise sponsoring Ola Hudson show. Ringo wearing rose colored glasses. One of many in pounding sun. I've been nervous all day. So much ambition so ill-spent. Show lacked imagination. Kim Milford flew across the yard on a wire, dangled then jumped onto mattresses. Tomorrow

Richard (Arons) returns from Brazil and I must leave the marble mansion with its spectacular views.

Monday rolled around and even though my neck and back were painful and spasming, I had no time to moan. I had gone to my doctor, but I needed to be alert. A neck brace helped, but painkillers were out of the question.

Monday, September 23, 1974

Robert (Blue) saw Chance and Loretta at Oingo Boingo Laurie Mann's party. Robert walked up to Lumilared Chance—"Did Linda Lane design your costume?" The designer turned green!

Now that we were down to the wire, friends were calling to ask what they could do to help, and artists were calling to ask if their work could be included in the event. Having to say yes or no to an artist made me very uncomfortable. I was all too familiar with the negative feelings associated with rejection, so I found it hard to foist that discomfort on a fellow creative.

Cecile was still ignoring Mike Van Horn's repeated requests for her to return his borrowed VW bug. The time for patience had passed. "Return my car now or I'll file a police report," was his final statement. Cecile returned it post haste.

CHAPTER 23

LUNACY

Rare as a unicorn, Donyale Luna sprang back into my life. Full blown, ethereal, and willowy as the thinnest aspen swaying in the breeze, with her huge, dark brown, almond-shaped eyes, she drew me into yet another adventure. Salvador Dalí called her, "the reincarnation of Nefertiti," a title she enjoyed personifying.

> Tuesday, September 24, 1974
>
> David Terry came by the shop. I was quietly frantic. Phone ringing, and no Edward to help me. Mike Van Horn hysterical over car as Cecile didn't leave it in front of the shop as promised. Had to pick up Ivan Dryer's photos. David wanted to go. While waiting at Laser Image I called the service. "Luna" and a local number. "Come to the Hollywood Bowl." We drove there and called. "I'll meet you at the street." David and I wait and try to find the goddess on Alta Loma, a terraced walkway street. David ascended a flight of stairs to search for Luna and met up with his business partner, Dan. We waited by crumbling white garages lined so as to look like a Roman gateway.

When I received the message from Luna I was elated. Filming her now would be a coup since the night of the fundraiser she would be in Rome, and having a video of her welcoming guests would set the tone for the evening.

I was living moment-to-moment, referring to my checklist every five minutes. Fate had tossed David and me together that day. Neither of us could have known that Luna was in town.

David Terry was short and ambitious. He was a photographer who knew Luna's work and desperately wanted to meet her, and even though I had another videographer in the wings, the moment seemed serendipitous.

The hunt for Donyale Luna amid the narrow, winding streets above the Hollywood Bowl was both exciting and exhausting. Like mice we scurried up and down constricting walkways, which were undoubtedly built during the Model T era.

David's business partner happened to live in one of the hillside houses and he was as anxious to meet Luna as David. "Let's call her once more," I insisted. "I've got too much to do. I don't have time for this." David countered, we called, and miracle of miracles, Luna picked up the phone.

"The black truck, Donyale. Don't you see the black truck?"

"No!"

We finally found each other on Hightower Road. David was overwhelmed, awed, as was everyone who met Luna for the first time. She instructed us to follow her through a maze of walkways until we reached her friend's house.

Suddenly, I was being put into a very awkward position. Videographer Paul Holman had volunteered to shoot the "Welcome to Creation" video, but I had David Terry and his insistent personality telling me that we should shoot the film at his studio now. Luna would be flying out later that night to join Fellini at a press conference in New York. It was either do it now or risk not doing it at all.

"Okay," I said, "Let's do it!" Once that was established, we gathered up Luna's bags and headed across town to David's studio in Culver City. As I recall it was a converted warehouse with high white walls, Plexiglass furniture, and lots of art.

Tall, willowy Luna wrapped herself in gold Lumilar and began moving sensually as she started performing. Her generous lips parted and "Welcome to Creation," softly slid out. She continued to invent a welcome speech as the camera rolled. It was as if David Terry had flipped her "on" switch. The model was making love to the camera and the cameraman was in full drool.

Luna had made me promise to get her to the airport for her 12:10 am flight, and as the time drew near, I corralled everyone into the car and off to LAX. Once there Luna realized that she'd left her portfolio at David's. The last flight was in forty minutes, so she changed her ticket to early morning, and back we tracked to Culver City.

Shooting resumed, only this time out came the drugs. Luna's flight wasn't until morning, and it was clear that the tall black model and the short white photographer could go on like this for days. Time meant nothing to either of them.

Next came still shots. David was having the time of his life, and Luna's poses were breathtaking.

Unfortunately, I didn't have the luxury of time. I had three-and-a-half days to pick up cups and plates and napkins and artwork and lasers and holograms. Having Donyale Luna literally welcoming everyone to the fundraiser would give Jupiter 5 a unique stamp of approval. David assured me her footage would play on a continuous loop.

I tried to take a nap—"tried" being the operative word. As Donyale's 8:45 departure time approached, we made sure she had her portfolio and giant bag before piling into the car. Luna, with inimitable Luna style, went into the bathroom and came out looking fresh and exquisite.

By the time we got going, it was after seven. Being strategic, we took residual streets to the freeway. "Stop," Luna called from the backseat. "I must have a Bird of Paradise." She gestured behind us. "There." She pointed at a thick grouping of Birds of Paradise next to a low chain-link fence.

I backed my rental car up in time to watch a sixtyish red-faced white man with a Santa Claus body, wearing a wife beater, navigate three steps and disappear inside the house.

"I must have a Bird of Paradise," she insisted, her breathy, urgent voice a command.

"We don't really have time, Donyale. We're late."

"I must have a flower!"

"Fine." The back door opened and Luna climbed out, gracefully ascending the concrete path and steps. She rang the man's doorbell and waited. Finally, he opened it.

"May I please have a flower?" She gestured towards the Birds of Paradise.

He was in shock. He disappeared and returned with a large butcher knife. Luna delicately took the lethal object, cut the flower, returned the knife and got back in the car, all while being filmed by David. Donyale had signed a release to me and me alone. I sensed that this was going to be an issue with the overzealous photographer.

We arrived at LAX. Luna had missed her flight by five minutes. Federico Fellini would have to hold his press conference without her.

Wednesday, September 25, 1974

Lloyd Cross coming down from San Francisco. All seems to be adding up to a great show. So excited to have Donyale video-tape and fashion shots—rounds out the evening—gives great plus for Jupiter 5. Feel really bad, ill, but can't stop for anything. Dropped off money to pay for food. Like a crazy person trying to attend to every detail and rechecking with Candy constantly. Rich Rallison will do laser show. Trying to reach Barry (Gott) to be certain he'll be there. So far all seems fine.

On Thursday David Terry and his business partner, Dan, wanted to see the venue. Once they approved we ventured down the beach to The Sand Castle for a drink. Things were going well until David handed me a bill for $104 for videotaping Luna plus $27 for looping. The fashion slides would cost extra.

"I didn't know it was going to cost this much. Paul Holman wasn't going to charge anything for shooting Luna."

"I'm a pro. Paul Holman's an amateur. You get what you pay for!"

David Terry knew I was stuck. He had the only footage of Luna, and I really needed it. The fundraiser was going to make or break Jupiter 5.

Friday, September 27, 1974

Josine Yanco-Starrels—Times—Cal State (she is an art museum director/curator). Chang Young Cho PSA 2:30. (LAX) Dropped Rodney at Fedco—picked up Chang, a gentle, child faced man of 29. Innocence stops when it comes to smoking dope. Did 1,000,000 errands—took Chang, Julia, Rodney & Edward to a Chinese restaurant for dinner. Came back and had Julia and Chang slice carrot sticks. Making carrot juice and dips to go inside frozen watermelon. All coming together, but hectic. No help from Cecile. Ruth racing to have program printed. Feel exhausted, neck throbs, back aches, but can't stop. Must make party a sweeping success.

I'm sure that everyone who spent time with Donyale has stories. And, I'm quite sure the prism through which she was viewed forms the tone of the recollection.

Luna will always be remembered for opening the couture modeling universe to women of color. She settled in Rome, married her longtime boyfriend, Italian photographer Luigi Cazzaniga, and had a daughter named Dream.

When Luna arrived in New York in her late teens, she neither drank nor did drugs. Sadly, the disease of addiction led to an accidental heroin overdose in 1979 at the tender age of thirty-three.

The following contract, drawn up September 25, 1974, memorializes events that unfold in this chapter. Time has encouraged the ink to fade. However, Luna's signature is as clear as the day she wrote it. Addressed to me, the document states:

RE: Videotape of Luna welcoming guests to Jupiter 5 party and fashion photographs wearing Lumilar and Jupiter 5 products.

I, Donyale Luna agree to allow Linda Lane to show the aforementioned videotape shot by David Terry and Dan Zimbaldi at the Jupiter 5 benefit. If there are further showings in which money or rental charge is involved, I must be notified and compensated.

The fashion photos are my gift to Linda Lane. If they are sold, the photographer and myself must be notified and compensation must be agreed upon.

CHAPTER 24

WELCOME TO CREATION

Saturday, September 28, 1974

Crazy day. Awoke with the birds to race to the party site. Julia organizing people. Spoke to a Valley radio station who plugged us. Chang (Young Cho) and Julia made Lumilar road signs. Candy picked up the liquor from Vendome and got an $80 rebate. Burt Gershfield brought a friend to help. Somehow all came together. Great food—yogurt dip for carrots and celery, chips, carrot juice, nuts. Halved out frozen watermelons with dips inside. Julia sang at 4 pm, music 7-8 Ronnie, Bud Cort sang, jazz pianist Joanne Graurer played the piano, Rocky played Lumilar congas, Dennis supervised hanging paintings. Edward did music tapes, Donyale's videotape was seen for a brief moment. Cecile, her daughter, Laurie, and Ruth arrived around 5 with programs. They were ready to socialize.

They were overwhelmed by the togetherness of the event. Candy spent the evening picking up cups. I wore an outrageous '50's dress with Fred Slatten pumps. Julie Christie, Margot Kidder, Danna and Ed Ruscha, the Harve Bennett's etc. were

> blown away. Lumilar car not there. Van Horn mad (at Cecile). Beth Ann Krier a no show. Patty Gilbert—Council for the Arts—seemed impressed with work. Celeste Fremont covered for Interview. I felt gratified. Produced a hit party. Candy was too busy picking up after people to enjoy herself. Dyan Cannon and Hal Ashby walked off with two bottles of champagne. Cecile took the spotlight being interviewed by Paul Holman. Judith Driver got Kirilian working around 11 pm. In spite of obstacles all was a much-approved event.

Malibu, like many upscale communities, included upscale trailer parks. Sitting high atop a hill overlooking rows of neat doublewides and the beautiful blue Pacific sat Paradise Cove's spacious clubhouse, the perfect setting for the Jupiter 5 fundraiser. A friend, and park resident, had reserved the venue for a mere $50, and that included a large onsite parking lot.

The day of reckoning had arrived and everything from food to paintings, sculptures, and holograms had to be in place by 4 pm. Dennis Pelletier came through, supervising the placement of artwork. Rocky arrived with his dazzling, Lumilared congas. Edward, who had taken over Dennis' shoe art, took care of music tapes. Dave Terry turned on the Donyale Luna "Welcome to Creation" videotape, although her voice was out of sync with the film. Ruth Tuska had the programs printed and delivered.

Most of my friends came through, working their tails off to pull everything together. The list of exhibitors that follows is from the printed Jupiter 5 program:

> "Welcome to Creation"—Donyale Luna welcomes you to an evening of contemporary magic. Videotape by David Terry—Studio 1817—Total media service.
>
> BARRY GOTT—"Fantastic Cities," sculptures exemplifying the Art of Technology. Plastics, lasers, circuitry. Dialogues on future phenomenon.

PAUL HOLMAN—Positive media in America. Videotapes, video feedback, electrokaleidoscope and taping.

"To Creation"—Slide show series by Dan Zimbaldi and David Terry—Studio 1817

RICHARD RALLISON—A collection of dichromate gelatin white light viewable reflection holograms. Laser light show: Deviations From Rectilinear Propagation.

POLI CECIL—The first artist to incorporate Lumilar into conventional art forms, thus innovating a new multidimensional art form: Lumilart.

AMIR FAÇADE (Dennis Pelletier)—Designed the first Lumilart car and other custom designs, including shoes for Fred Slatten.

CHANG YOUNG CHO—Contemporary painter, lithographer, Lumilartist.

MICHAEL KORHONEN—Tantric dance, auric portraits.

ROBERT WILSON (it should have been Nelson)—Designer of Jupiter 5 invitations; California Exposition and other art show awards for fabric designs. Invitation to exhibit in contemporary Crafts of Americas Show in Denver and South America. Graphic designer now creating a line of Lumilart Christmas cards for Jupiter 5.

IVAN DRYER—Artist exhibiting a collection of laser photographs. Mr. Dryer is well known for "Laserium," currently showing at the Griffith Park Planetarium.

MICHAEL FOSTER—Jeff Allen presents a collection of Michael Foster's printed holograms including the first holographic record master and a 3' x 5' diffraction grating panel.

School of Holography—MULTIPLEX COMPANY, San Francisco, California. Lloyd Cross and his associates exhibit the first multiplex holograms and white light viewable rainbow holograms.

DR. RALPH WUERKER—Holograms created by Dr. Ralph Wuerker for TRW. First art artifacts recorded by holography.

MIKE LEVIN, DAVIS BEYERLE—Electric sight, special effects and lasers.

RUTH TUSKA—Auric portraits of Sam Peckinpah, John Wayne: Lumilart star portraits of Basil Rathbone, Joan Crawford, the Marx Brothers. Artist, writer, businesswoman, Co-founder and Managing Editor of Views and Reviews Quarterly Magazine of the Reproduced Arts.

By late afternoon the stage was well on its way to being set. I was convinced that this would be a historic night for holography, that the artists and holographers who participated in the benefit would never find themselves together under one roof again. Forty-five years later, my prediction was correct. Ivan Dryer, Lloyd Cross, and Richard Rallison are all remembered for their pioneering accomplishments with lasers and holography. Dr. Ralph Wuerker is known as a pioneer of laser physics and holography, while Michael Foster has been called the "Howard Hughes of holography"—the mysterious genius inventor.

"Welcome to Creation" was indeed a you-had-to-be-there celebration of 1974, art, science, and Hollywood. I was listed as the producer; Candy Clark, the chairwoman. The technical producer was HCCA and Ruth Tuska. The list of special thanks included over forty individual names and companies. Clothing designer James Reva and his partner Mark Jarrett, Lou Adler and Ode Records, MCA Records, Avco Cinema Center, The Luau Restaurant, Dan Tana's Restaurant, and Universal Tours are but a few of those listed as having contributed donations.

The beginning of the evening was anticipatory and civilized, but as the guests and exhibitors tucked into bottles of champagne and carrot sticks, things became a little unhinged. Bud Cort, who you may remember from *Harold and Maude*, sang. And not well. Joanne Grauer, a composer and singer with a new album, entertained. Rocky, looking like African royalty, played his customized congas.

Since Donyale Luna's voice was out of sync with the film, I considered it a total loss. My anger and disappointment were palpable, because now there was nothing to fascinate guests while they checked in. The volunteers manning the welcome table got restless, and, to put it bluntly, no one was minding the store after the initial onslaught. I can't say whether it was the mindset of the day or people just behaving badly, but many of the guests arrived demanding free admission.

Fueled by champagne and live music, the fundraiser quickly became a fabulous evening of controlled chaos. Guests who had never seen a hologram before were awestruck, and nothing made me happier than watching Hollywood's movers and shakers discovering otherworldly 3-D magic. As promised, Candy Clark had delivered A-list guests who appeared to marvel at the future.

Margot Kidder, actress Dyan Cannon, director Hal Ashby, and producer Si Litvinoff could be seen talking to exhibitors, taking it all in. Everything was going according to plan until I realized that my friend, *Los Angeles Times* writer Beth Ann Krier, was a no show. Lester Sloan left me a message saying that he was stuck in a hotel room in Long Beach waiting for Richard Nixon to leave the hospital. So no *Newsweek* either. *Interview* magazine came through with Celeste Fremont and the stipulation that we supply the celebrity photos.

Once the dust settled, Candy and I realized that we had put on a memorable event, but an event without the manpower needed to maximize fundraising. Instead of socializing, Candy spent the better part of the night picking up after people, and I zoomed from one exhibit to the next juggling hostess duties with catering oversight.

When we ran out of champagne, I foolishly sent someone to buy more. The Kirilian photography exhibit didn't start working until 11 pm, and by the end of the night, I was just happy it was over. A few holograms, a Chang Young Cho painting, and a couple of Barry Gott's sculptures were sold.

Sunday, September 29, 1974

Try anti-climax to end all anti-climax'. Depression. Stopped payment on David Terry's check.

By Monday the results were in: "Welcome to Creation" was a social success, and a financial disaster. The costs were too high and the do-it-yourself money-saving measures led to volunteer burn out. I can remember my own feelings of stress occasionally mixed with exhilaration and sometimes even joy.

Cecile, Ruth, helpers Edward and Rod, and I met at the store to explore Jupiter 5's future. Cecile aired her grievances and I aired mine. We could pay the rent, but we didn't have enough in reserve to give anyone a salary or a sense of confidence.

The phone number on the Jupiter 5 program was wrong, off by one digit.

The professional pictures shot by an amateur turned out to be unusable—for example, Julie Christie's face with Hal Ashby's back. There were no celebrity shots acceptable to the *Interview* editor. And there were a lot of good photographs of people I didn't know and couldn't identify. We didn't get the publicity we needed, but at least we planted the seeds of recognition in the Hollywood community.

A year and a half later I found myself at Universal Studios to discuss using holograms for *Close Encounters of the Third Kind.* When I was introduced to the film's producers, Michael and Julia Phillips, Julia saw the multiplex cylinder I was holding and quipped, "I know all about holograms. I went to a big party in Malibu." Then we spoke briefly about "Welcome to Creation."

Julia Chuhralya had finally had enough of Richard Arons. She was moving out and he was furious. He wanted to make her a star, and now she was throwing it all away for a bisexual Lothario named Creed.

Paula Cecil was in London for two weeks, leaving Dennis in charge. And take charge he did. He invited Julia to stay at the house while she sorted things out, and I was invited back into the fold. Gleefully Dennis and I restarted our good, Hawaii-happy relationship. For me it was an oasis of much needed peace and serenity.

Wednesday, October 2, 1974

My right eye has atrophied or something. It doesn't push into the socket as it did a year ago. It feels odd.

Candy and I did a righteous thing by giving the party. Artists are getting work, new projects are being generated. All is moving in the right direction. Thank God!

Jupiter 5 was located in Santa Monica at 1843 Lincoln Boulevard, a main thoroughfare lined with commercial businesses and lightly peppered with retail shops. The rent was cheap, but, that said, unless a pedestrian or a driver was specifically looking for holograms, they weren't going to find us. There was little, if any, foot traffic.

The fundraiser solidified the schism between Cecile and myself. We'd been at odds for a long time, Cecile blaming me for the fire and zero insurance money, neither of which I had any control over. I blamed her for not being a team player, for disappearing when she was needed the most, and for not returning the Lumilart VW bug when promised. Her lack of participation in the benefit as well as giving her store key to "a friend" seemed to be her way of silently shouting, I'm outta here!

On Friday I spent from 1:30 to 4 pm at the doctor's office. Dr. Sellman felt the knotted muscles on the left side of my neck and my lower right shoulder. He ordered X-rays, which led to hours of heat, traction, and massage therapy. The car accident coupled with extreme stress had taken a toll on my health.

The doctor's appointment made me late for a meeting with Cecile and Ruth, and they couldn't wait to castigate me. Ruth had been a successful businesswoman. She knew what we needed—regular business hours, salaried employees, and a business plan. She told us that certain stringent measures had to be taken and that she was the one to take them—and, if we didn't cooperate, Cecile and I were doomed.

And I believed her.

Friday, October 4, 1974

I fell asleep during Highlights From Cabaret. Larry (Becker) saying, "Good night," and "Oh, did you meet John Avildsen. He directed Save the Tiger." Wham and I'm into Revelation II potential and he's interested in reading it, and on it goes. He's charming and nice and I feel a bit too dogmatic and full of myself and I apologize.

Saturday I worked at Jupiter 5 all day, selling a few feet of Lumilar and trading a woman a couple packages of stick-ons for a planter. I kept thinking about John Avildsen, the director with the sharp, dazzling blue eyes. Meeting him after the screening had been a breath of fresh air. Suddenly, in the midst of hopelessness, I felt a tinge of optimism.

After all, I'd read about Thomas Edison, whose teachers had told him he wasn't smart enough to learn anything. He went on to hold a thousand patents. Theodor Geisel, AKA Dr. Seuss, wrote his first children's book, *And to Think That I Saw It on Mulberry Street* in 1937, and had it rejected by between twenty and forty publishers. Even Walt Disney praised "having a good hard failure when you're young."

Everyone I'd read about shared one common message: *Never give up!* Grit, persistence, and certitude would eventually put me in the right place, at the right time for a successful outcome. With this in mind, I went to the shop.

Sunday, October 6, 1974

Left Alan's—went to the shop to clean. Was thinking about John Avildsen so I called him and he said, "I just tried you and your line was busy. You saved me a nickel." He came to the shop and I explained holography while he came up with ideas for a holographic motion picture. He has magical blue, effervescent eyes.

"Would you like to take a drive?"

"Yes."

We drove to Malibu to see some of John's friends from New York. Bo Goldman. We were met at the door by the nanny who extended her hand and introduced herself. William Goldman is off to Switzerland and I have been privileged to have a family dinner with Mrs. Goldman and the children—extraordinary. John fascinating.

The rollercoaster just kept taking me on the ride of a lifetime. Every time something fell apart, I turned around and another door was opening. This time I felt privileged to show John Avildsen different kinds of holograms, and to brainstorm about their potential use in movies and advertising. He'd begun his career as a cinematographer so he quickly grasped the concept of transforming motion picture film into three-dimensional imagery.

We clicked. We were on the same page. When we discussed *Revelation II* and a larger-than-life-size hologram of Christ floating in front of a movie screen for up to thirty seconds, the wheels of his mind started racing. Could it really be done? Yes!

On Monday I arrived at the shop to find Ruth madly working a calculator. The two of us were in mid-vent about Cecile vanishing to Mexico with Kirilian photographer Judith Driver when the door chimes sounded, and a flamboyant nymphet wearing just enough, with a Lumilar star in the middle of her forehead, breezed in and announced, "I'm a stripper. And I have Lumilar pasties!"

She had our undivided attention. "Where," we asked, "did you get these fabulous Lumilar stars and pasties…?"

"I went to a party at Julie Andrews'—a hundred dollars per person. They had a circus. It was wonderful. They gave everyone a Lumilar star."

It seemed that she'd attended this party with Star Rainbow and the Laguna Beach airbrush artists, and the only common thread that I could think of with access to rolls of diffraction grating patterns was Cecile. Suddenly our schism became a giant sinkhole.

> Tuesday, October 8, 1974
>
> Every time I get out of (physical) therapy I feel worse—headaches and tension. Accident makes it impossible to concentrate on the shop. Valiums kill anxiety as well as alertness. Don't know what's going on.

I was having physical therapy five times a week. I'm sure a psychiatrist would have been helpful too.

One thing that Cecile and I were both passionate about was making the phone company add "Holography" to the yellow pages. We pestered and insisted until they finally acquiesced. This small victory led to a man finding Jupiter 5 and calling to order a holographic prototype of a boat in a bottle. He came to the store and took Ruth and me to dinner and was so impressed that he offered us his office space gratis until we got off the ground. Of course, this was contingent upon him receiving a successful prototype.

Richard Rallison was the holographer for the job. If he could manufacture a good dichromate hologram of this man's cherished ship inside of a bottle, a commercial baseline could be established. Rich would be able to quit his job at Hughes Aircraft and make holograms full time. Jupiter 5 would receive a finder's fee, and in the process, gain a financial backer.

Then, boom! Rallison's kitchen blew up. He survived with a few burns, but the kitchen itself and the client's childhood heirloom were reduced to shards of glass and wood. There would be no more dichromate holograms of any description until Rich's kitchen was rebuilt. I also seem to remember his landlord telling him to find another place to make holograms.

Wednesday, October 9, 1974

Ruth and I discussed incorporating Jupiter 5 sans Cecile. I went to see Fred (Butler) who offered to get me a house. He apologized for his macho actions and laughed at black Richard's taking me for con supreme. Motown is supposedly picking up his (Fred's) tab, but no contracts and he and Al (his song writing partner) could have all the money (advanced) deducted from their royalties.

Seeing Fred Butler forced me to analyze my feelings. Why had I rejected Fred in favor of con artist Richard Aaron? And was Richard Aaron even the man's real name? Then I realized what was at the heart of the matter: my sense of freedom. Fred would have been a really good partner had I wanted to settle down, but he wasn't the kind of man who would tolerate a woman striking out on her own whenever and wherever she pleased. I needed a man who understood both sides of the desk; an artist with an equal measure of business acumen.

John Avildsen was busy finishing *W.W. and the Dixie Dancekings* starring Burt Reynolds. He ordered a reflection hologram of a Sunco car for his young son and I delivered it. He showed me home movies—his shake shingle house at Trancas and shots of friends. He was fascinated by lasers and the cinematic magic that could be achieved using new technology. Both of us were under a lot of stress, he with *Dancekings*, and me with my whole life. Spending a little time with a kindred spirit gave both of us a much-needed breath of fresh air.

Saturday, October 12, 1974

Wanted to impress John (Avildsen) so sewed sequins on red cape and looked as good as possible. Candy (Clark) and I went to a recording session for Henry Kissinger's album at the Record Plant. Dashed to 20th to see a screening of John's film, W.W. and The Dixie Dancekings. Burt Reynolds drives into an "SOS" gas station and robs it. He's a con man.

It had been a fun day, and I couldn't wait to see John's film. I'd invited Candy Clark to come to the studio screening. The lights dimmed, and charming, toothy Burt Reynolds robbed an SOS gas station. The more of the film I watched the more alarmed I became. My script, *Revelation II*, is the story of an artist/con man who sells the Second Coming to a black evangelical preacher who happens to be the founder of Sacred Order of the Spirit—SOS. *Revelation II* is set in the South. *W.W. and the Dixie Dancekings* is set in the South. I felt my world shrivel. There were so many coincidences. Why hadn't John said something about SOS? We'd discussed the premise of my story!

In my Valiumed haze I fell apart. I felt as if someone had read my screenplay and decided that "SOS" was a great name for a gas station. It never occurred to me that there might be real SOS gas stations. Why would anyone want to make my script after seeing John's movie? Plus, his film was scheduled for an Easter 1975 release, and my film, when it was made, would need an Easter release.

Looking back I now realize I was burned out, overly emotional, and definitely not thinking clearly. It took forty-five years for me to revisit this unfortunate memory to realize how broken I was. My hysteria spilled over to John, who was dealing with his own director's issues.

> Sunday, October 13, 1974
>
> Devastated, feeling rejected, dejected, a mess! Stopped eating. Feel like a cow. Wonder why so much that's negative must go down. Economy in bad shape for pioneering efforts. Holly Harp did 50% of projected sales for the year. Shows that even #1 is off by half.

My agent spoke to John about *Revelation II* and the references to SOS. The director assured him that Thomas Rickman, the screenwriter of *W.W. and the Dixie Dancekings,* knew nothing about my script. That, in point of fact, there were SOS gas stations operating in the South. I also spoke to John briefly before he left for the airport and New York. If he didn't like Hollywood before, he really didn't like it now. I had taken a warm, wonderful connection and blown it to smithereens.

In 1975 John Avildsen began directing a little film called *Rocky.* Released in 1976, it became an international blockbuster. The following year he won the Oscar for Best Director. In all, *Rocky* received ten Oscar nominations, winning three, including Best Picture.

CHAPTER 25

BEGIN AGAIN

Monday, October 21, 1974

Dr. Sellman prescribed a double dose of Valium to allow me to cope.

Holographers had begun to splinter into three groups. First, there were the physicists who were pioneering laser photography at universities, government laboratories, and large corporations. They had funding for powerful lasers and expensive Agva plates. Most of them were working on information storage and finding finite flaws in technical equipment.

One notable exception was Dr. Stephen Benton at Polaroid. Edwin Land, the inventor of the Polaroid camera, wanted to fabricate a camera that shot 3-D images. Land took young Benton under his wing and allowed him to use his scientific brilliance and artist's eye to make a historic breakthrough: the rainbow hologram. Heretofore holograms had to be illuminated by a laser. Now they could be seen using something as simple as an ordinary spotlight.

Next came the breakaways—the Richard Rallisons and Michael Fosters. They wanted to mass-produce holograms. As early entrepreneurs, they risked it all for a chance to gain a foothold and establish themselves as frontrunners.

The fallout from Richard Rallison's kitchen explosion, both physically and professionally, left the holographer twisting in the wind. He would have to keep his day job at Hughes a little while longer.

Rich's mentor and fellow Utah holographer Michael Foster, had just suffered a disappointment when Columbia Records canceled production of his holographic record center.

"Mike's angry," Rallison told me. "He doesn't want to hear the word holography anymore." Foster's front man, Jeff Allen, was known to be talking to two other holographers about representation. Holography had become a nuanced business starring a small collection of very bright, very stubborn science geeks who, even without the Internet, were making global connections.

Artists made up the third group. They may have come from science or fine art, or simply become smitten upon seeing a hologram. Anait Stevens is a perfect example of a woman who got hooked on the medium, hired holographers to teach her how to make quality holograms, and ended up with her three-dimensional compositions in museum collections.

In 1974 holography was in a nascent stage. I could envision where it was going, but not being a physicist I was on the outside looking in. It was my passion, but I had to be practical and do something that would mediate anxiety and pay the bills.

Tuesday, October 22, 1974

I went to Kim's (Milford) ABC taping. Julia still cold. Creed warm. Kim a definite superstar. Ruth (Tuska) arrived with motorhome sans children. Has it together. Spoke to Ron (Suppa, my lawyer). He loved Paris—thinks until the books are in order there's no way to make a final settlement with Cecile. She came to the shop and took liquid crystals and a light fixture then wrote a nasty note asking me to sign her drafted agreement. Offensive move—strange—off the wall from previous discussions. She knows I won't sign anything without Ron's approval or when Mercury is retrograde.

Malibu residents had always been receptive to holograms, so it made sense to spend a weekend participating in a parking lot craft fair. It was at this serendipitous moment that I met a beautiful young couple who lived down the road. I'll call her Cassie, and her young lover, Gino.

Cassie was a thirty-two-year-old painter who had parlayed her Scarlet O'Hara beauty and charm into a very sharp, albeit, creative real estate deal with her estranged husband.

Gino was in his early twenties, a tall, handsome puppy ready to do his mistress' bidding.

The couple enthusiastically described a sprawling Point Dume compound on Cliffside Drive with private bedrooms and communal spaces—an artist's colony overlooking the Pacific. Cassie described the property beginning with the four-car garage. The second story was a complete apartment, and downstairs, the space at the far left had been converted into a private bedroom.

"I only have one bedroom available," she said, smiling. "Come by. Take a look. If you want it, you have to act fast—and—I'll need a deposit."

As soon as the fair wrapped, I made a beeline for Cliffside Drive, where large, secluded homes lined both sides of the street. Of course, the ones on the water like Cassie's were the most desirable.

All I could see from the street was the two-story garage. I parked and began trekking through a half-acre of overgrown garden until I saw the sprawling, white, single-story Moroccan-style structure on the edge of the cliff. It looked new, finished and yet unfinished. Just thinking about it brings back memories of the spectacular ocean view and the luxurious lifestyle that I imagined went with it.

As I quickly discovered, Cassie was not just beautiful, she was also cunning. She'd talked her conservative husband into funding her real estate pursuit, convincing him that her brilliance was their brilliance. Maintaining the property and keeping up with her Malibu social life didn't come cheap, so she'd wasted no time renting the private bedrooms to young, hip tenants.

Cassie's private domain was a large bedroom/art studio that wrapped around the right side of the main house. On my "welcome to my home" tour, she explained that renters had use of the living room, kitchen, grounds, and private Point Dume beach.

The two bedrooms that were attached to the main house were private, self-contained, and could only be accessed by a dedicated key.

The moment Cassie unlocked my bedroom door and I saw the white walls textured with fist-sized rocks, the shiny king-size brass bed with crisp white linens and a down comforter, and the marble bathroom with two sinks and a bathtub large enough for a rock band, I was all in.

I was tired of being a gypsy. The property had the freewheeling spirit of an upscale writer's retreat. A deal was struck—all of this for $300 a month—and I had to get the money together. I could think of little else, and I couldn't have been happier.

> Thursday, October 24, 1974
>
> So sick of physical therapy five days a week. Rodney agreed to run the shop. Ruth agreed to assist with no strings. Julia is fine and saying, "Linda, go back to writing."

Somehow, whenever I was at a crossroad, I turned to Shelly Davis, who was only too happy to welcome me back to his office at MGM. He and his partner, Shelly Brodsky, had a studio deal that was based on them bringing fresh material—TV pilots, movies-of-the-week, film concepts—to MGM executives for a first look.

The two Shellys were clever, but as it turned out, not always lucky, and luck plays a major role in getting projects off the ground. Ernestine, their studio-assigned secretary, had been let go. They could no longer afford that luxury, so Shelly Davis came up with a solution: me.

Pointing at the front desk with a nice Selectric typewriter, he said, "Linda, just like before, you get a third of everything we develop together. Here's your desk, your typewriter. We don't care if you work on your own projects. All you have to do is answer the phone, and a little typing…". He smiled as if he'd just handed me front row tickets to the Beatles.

What this translated to was secretarial work in exchange for a desk at MGM, and the possibility of future deals. In the business it's known as an "if come deal." If it comes to fruition, you'll make money, and if not…oh, well, nothing ventured, nothing gained.

I believed in the two Shellys. They were very supportive of my work. My strength as a writer had always been coming up with original material. I had a vivid imagination, and I loved to incorporate facts with fantasy. For example, in *Laser Lady*, my original screenplay, I described a secret cloning spa where the politicians who voted to make human cloning illegal were the very ones paying to have their own clones kept alive—just in case they needed a spare, compatible body part.

For me, writing has always been a solitary occupation. I needed a pleasant setting with solitude, healthy snacks, hot tea (herbal and caffeinated) a typewriter, and reams of white paper. An artist's colony on the water with like-minded artisans would allow me to calm down and gain a renewed sense of self. The road was awash with positive signs.

Unfortunately, I was still seeking my father's approval. As a child I idolized him. After all, he was California State Skeet Champion, a world record fisherman, and the owner of Field Trial champion Springer Spaniels. He had completely renovated the Egyptian Theater in Hollywood and built new movie theaters. He was a director of the Associated General Contractors and served as a Trustee of the Carpenters Union. He was appointed to a post by Governors Brown and Reagan. He had a very long list of accomplishments, and I was determined to be counted amongst them.

This, however, was easier said than done. My mother had written a screenplay and several books, none of which were sold or published. Instead of being supportive my father belittled her: "If you were any good, they'd pay you!" He measured success in terms of financial remuneration. If you were paid, you had talent, and if you weren't, you were to be pitied.

Saturday, October 26, 1974

By hook or crook I want to move to Point Dume. The cast are all stars: (Cassie) is a love, Liz Taylor-fresh and always blossoming. It's no wonder she owns a lovely home at such an early age. At any age.

Gino, her love, her lover, her tidal wave of passion—as passionate as all of her paintings. So in love while killing each other.

Cecile came by the shop—discussed Jupiter 5. "Well, if you want to junk it, I'll take it on." "You will?!?" "Yes," she said, and so we agreed that as of January 1975 Jupiter 5 will be Cecile's operation. I'll give up control, take a percentage of gross and go back to writing. Rodney will receive a percentage and run the store.

I began going to MGM and writing my own material. I came up with an idea for Lily Tomlin that was sent to her manager. I came up with a visual concept utilizing lasers for Kim Milford's new television show. Some days I was an unpaid secretary, while on others, when the producers were out of the office, I was off and running my own creative race.

Saturday, November 2, 1974
Halloween Party—Bill MacDonald's

What a combination of fame and fortune hunters mixed into one. Roman Polanski wearing top hat and tails, black mask—blank behind his eyes, a well of emptiness shielding him from future disappointments. Zigi, his girlfriend wore a chic wool jersey dress and a coat of black feathers. She took it off—it was moved and she freaked.

"Where's my coat?!? I paid $2,000 for it!" I pointed to a corner saying, "I hope it doesn't lay an egg."

An incredible party. Bill MacDonald on crutches and wearing a crash helmet. David Stein, i.e. (handler for) Bernie Cornfeld's girls, David in black suit and white tie. Jules, a shop owner so Mafia dressed and acting I thought he stepped out of the Godfather. 150 costumed characters rocking in and out of each other's lives.

Bill MacDonald aspired to be Jay Gatsby. He had the perfect party house situated on top of a mountain with a 360-degree view of Los Angeles. On very clear days Catalina Island could be seen on the distant horizon. I especially enjoyed his bacchanalian gatherings because I was allowed to morph into Lady Lumilar or Laser Lady, as some preferred to call me.

On this Halloween I went as a flapper wearing a see-through vintage, long black silk chiffon gown with black fringe covering the parts that would leave nothing to the imagination. And, of course, I wore my light reflective six-inch Fred Slatten pumps, and prismatic diffraction grating bracelets.

Escaping the loud music, I meandered outside and down the sloping green lawn to lose myself in the twinkling lights of the city. I hadn't been standing there very long when I noticed a man wearing a foppish, seventeenth-century costume. He introduced himself as Guy Fawkes. I found his English accent and self-assured demeanor intriguing.

As we spoke I discovered that he was an artist named Paul Whitehead, famous for having designed Genesis' iconic album covers. Guy Fawkes, I later learned, was a radical Catholic who in 1605 was arrested in England while guarding explosives set to blow up the House of Lords, and with it, King James I. Never having traveled to England, and not being up on British history, I didn't know anything about Guy Fawkes, and my lack of sophistication was not a plus. Paul had studied Fine Art at Oxford, and he didn't suffer fools. I did, however, peak his interest when I told him about holograms in relation to Salvador Dalí.

We exchanged phone numbers, and I invited him to my new Point Dume artist's compound for an unconventional housewarming party. No gifts, just baptism by champagne.

When that day came around, Beth Ann Krier brought another writer from the *L.A. Times*, and Candy Clark brought her brother, Randy. Ron Suppa, my lawyer, fresh from a trip to Paris, and clothing designer Jim Reva, watched Yarousha's film on Tibet. Artist Robert Nelson showed us kimonos that he was taking to Sotheby's. In all, between my friends, Cassie's friends, and friends of the other boarders,

I made chicken curry for fifty people, and, in doing so, launched myself into my new salon lifestyle.

> Tuesday, November 5, 1974
>
> 1779 Orange Grove—Kim Milford—Laserium
>
> Kim (Milford) told me the premise for his new series, and I came up with an idea—add Laser Lady—must write it up. Took Kim to meet Ivan Dryer and see Laserium show. Spectacular—he added a krypton (laser). Kim interested in hiring Laser Images for his tour.

Having observed a number of Wendell and Alice Corey's elite Hollywood parties with the Robert Mitchums, the Gregory Pecks, and the Richard Widmarks, I had developed a salon sensibility. Writers, directors, producers, playwrights—East Coast, West Coast—there was always an eclectic group of show business friends and acquaintances discussing everything from their children to their next project.

That was the life I wanted. I wanted to inspire and be inspired. I wanted to surround myself with creative individuals, especially those who shared my passion for lasers, holograms, and film. I felt that I was entering a new chapter in my life.

> Thursday, November 7, 1974
>
> Cecile took shoes back to Fred Slatten, and they deducted $ for my shoes. Cecile really mad. Wanted to get Jupiter 5 together, but is rapidly growing disenchanted. I must act fast or she'll change her mind. Lumilar (rolls) getting low... She realizes the difficulty of such a business, and is having 2nd thoughts. Phone bill $137—many calls to NY, and to NY from San Francisco and Washington DC.

Jupiter 5 was on life support. It was doubtful that anyone could make it profitable without a serious infusion of funds, and I didn't spot any white knights on the horizon. If I was going to survive, I would have to concentrate on writing, and, when time and social contacts permitted, agent holographic deals.

> Friday, November 8, 1974, 10 am therapy (physical)
>
> I've really moved into my castle by the sea. It fulfills every dream—brass bed, sunken bathtub. Home. Cassie is a love, an angel. Alfred is a great man though only 25. Feel so depressed at the studio. Candy (Clark) came for lunch and drew Danny Melnick (studio head) to our table. Shelly thinks she's right for The All American Girl. Shelly still doesn't understand where I am— head wise. He thinks I should be a robot, sit down and work, work, work! He thinks I'm a socialite. And, in a sense, I do have that side.

My neck was still in pain, but living and working at home in Malibu gave me relief. Shelly Davis had come up with an unusual idea for a film. On April 11, 1926 a man named Edward W. Browning, aged fifty-one, had married a fifteen-year-old schoolgirl named Frances Heenan, aka Peaches. The headline on the front page of the *New York Times* read:

> **BROWNING, 51, WEDS HEENAN GIRL OF 15;**
> **HER PARENTS ASSENT.**
> **Ceremony in Taxi Garage at Cold Spring Where He Leased Two-Acre Estate.**

Beneath the subheading in all caps it read: "MOTHER AND FATHER THERE." After more subtitles, in smaller print it read: "Wealthy Realty Dealer Divorced First Wife—Had Vainly Tried to Adopt a 'Cinderella'."

There were lots of lurid details with only one conclusion: Daddy was a pedophile, and Peaches and her parents were gold diggers. These were not exactly characters you wanted to root for, but the story was so bizarre, and Shelly's enthusiasm so contagious, I convinced myself that this was just crazy enough to be produced.

One day while eating lunch at the studio commissary, Shelly and I spotted Walter Matthau.

"Go over there and introduce yourself," he instructed. "Tell him you're writing a screenplay that's tailor-made for him. Go pitch Daddy and Peaches."

At his urging I walked across the large, noisy room and over to Mr. Matthau where, upon mustering my courage, I politely introduced myself.

"Have you ever heard of Peaches and Daddy?" I inquired.

He had not, so I explained the historical context, assuring him that we were writing a period comedy.

The star listened and asked if there was a completed script. "No," I said. To which he shrugged and said he couldn't give an opinion without having read the screenplay. He then offered to read *Daddy and Peaches* when it was finished. I was young, and he was very kind.

This is another one of those instances when I have to ask myself: What was I thinking?! It doesn't take much to know that there's nothing remotely funny about a fifty-one-year-old millionaire who attempts to adopt underage girls, and when that fails, marries one.

There were always rumors in Hollywood about producers who preferred jailbait, as they called it. Whether it was true or not, they would have been the last ones to green-light a picture about an obvious pedophile. I now owned one-third of a project that no one would ever touch.

Monday, November 11, 1974,MGM & therapy (PT)

Worked like crazy to finish Peaches & Daddy—almost tricked by Shelly's "We could go see Link and Levinson today..." (big television producers) so I get on the typewriter and whack away and come up with half a finished treatment. Shelly brings me lunch. Went home exhilarated. Physical therapy helping.

The house is like a nut house. William (my astrologer) called Cassie for a reading tomorrow—he never does that. Something is really wrong. Mary and Jay are telling "their" side—how Jay lost

> $50 worth of grass and Alfred is the only suspect. Is it Alfred? Is he a total con artist, having a Christ-like façade and enough cool and mind game power to casually destroy a household?

I had barely settled into my idyllic new home when all hell broke loose. My narcissistic housemates were at each other's throats over a lost, stolen, or misplaced baggy of marijuana. Blond-haired, blue-eyed, all-American Jay did commercials and tried to get acting jobs. His girlfriend, Mary, who hadn't yet graduated from high school and saw no reason to get a GED, aspired to become a model. She told us her mother was a high-priced Las Vegas call girl, and her father was a marine biologist. Neither one of them seemed to take any interest in her. Fortunately, Mary had been blessed with long legs and a Mona Lisa smile. Her brown eyes lit up at the sight of gold wherever it glittered. The couple spent most days on our private beach.

Being English with wispy brown shoulder length hair, wearing caftans and leather sandals, Alfred exuded calm erudition. Who should we believe—the blond-haired surfer or the brown-haired philosopher? Or neither? Maybe Mary was the culprit. Or maybe Cassie hid something so well she couldn't find it. It didn't take long for my dream house to become another version of the circus I'd left behind.

> Wednesday, November 13, 1974
>
> Finished 18-page "The Laughingly Lurid Loves of Peaches and Daddy" treatment. Hyper-disillusionment—Jupiter 5. Can finish all paperwork by Friday. It takes hours of pain to recount one year's experience. So many rip off artists. So many millionaires waiting to invest and nothing concrete materializing for many reasons.

Cecile had taken over Jupiter 5. I was out. That much had been agreed upon. I, however, still had projects in the works. Artist Robert Nelson had designed an intricate snowflake pattern that was going to be printed on Lumilar for Christmas cards. The light-reflective material would then be glued onto heavy white card stock. On the back it would credit "Hologram Press."

I needed diffraction grating material to send to the printer, and Cecile took a hard line. Either I paid for it up front or I didn't get it. She also refused to give me my Reinell commission or a Chang Young Cho painting commission.

I took a leap of faith, paid for the material, had the black snowflake printed on prismatic silver foil, and realized a tiny Christmas miracle. Selling for one dollar a card, which was high then, the series sold out instantly.

Meanwhile, Shelly Davis hatched a new plan. His friend, Bernie Deutsch, a gnome-like Jewish financier and clothing manufacturer from New York, wanted to discuss putting Lumilar on his ready-to-wear line. I still had a relationship with the Diffraction Company so I thought it was worth pursuing on my own.

If I'm not mistaken, Bernie was putting money into Now Productions, Shelly's production company. This gave the producer an idea. If he could get his financier friend to invest in some of my holographic products, I'd have enough money to pay my rent.

Shelly arranged a luncheon meeting at the studio commissary, and by the end of the meal Bernie was almost giddy with anticipation. He wanted to see every hologram in my arsenal.

This was Shelly's way of letting himself off the hook. On the face of it, it seemed counterintuitive. Then, upon closer scrutiny, I realized that, come what may, he saw it as his way of rationalizing my unpaid writing and secretarial services. He thought he was doing both of us a big favor.

I invited Bernie and his assistant out to Malibu to see the line I'd put together, as well as the Lumilar dog and cat collars Cassie was making. Bernie's assistant was excited about adding Lumilar to their ready-to-wear line.

Everything was going well until Bernie told me he needed to manage me for 25 percent. That meant one-fourth of everything I'd earn from writing and holography. He wasn't really in show business, although he wanted to be. He was based in New York, where he manufactured clothing. I wasn't sure what Bernie Deutsch could do to help me? Was he going to fund my projects? Raise money to fund them? Hope that something I wrote or created would hit the jackpot? I needed a real paying job as opposed to a would-be handler giving me directions. I had finally gleaned enough common sense to say a resounding NO to Bernie's offer.

> Tuesday, November 19, 1974
>
> Alfred hasn't paid rent in two months. Jay came in to invite me to a dinner party at Michael Butler's in Montecito. Got all hyped. Neck still hurts. I can't seem to get it all together. Working against time. Lumilar still the most exciting medium. Bernie called from New York to say hello. House menagerie continues. Gino hanging on my back window at 4 am because he's mad about my taking Cassie to a party where she might meet a new, interesting, substantial man. If Peter (Cookson) hadn't sent me $100 I don't know what I'd do.
>
> On the way home stopped at Brentwood Fotomat. A girl looks at (my) car—"Oh, you have Lumilar, too." I smiled and said, "Yes," to the stranger.

Beyond the revelry and chaos at Cassie's, which included a continuous stream of people, many of whom stopped by for the key to the private beach; some with children, some with children and dogs. There was seldom a dull moment.

Once again my father allowed me to work at his office so that I could pay my rent. I continued going to MGM every chance I got, and I made peace with the fact that Jupiter 5 had closed its doors. Cecile moved the bank account, and now there was only one signer, Cecile Ruchin. Edward and Rodney, the boys who had worked so long without a paycheck, still asked me when the insurance money was coming.

I kept telling them, "There is no insurance money. It wasn't my policy. It was my father's, and he thinks it's my fault that someone set the fire. And I think he's right. I think we know who did it."

Neither the fire department nor the police cared to spend time on my father's office fire. It happened late on a Sunday night, and someone called the fire department. No one died, and only a small portion of the building was seriously damaged. It just so happened that that small portion held my most precious possessions along with the future of Jupiter 5.

Thank goodness the year was almost over. Living in Malibu I felt like I was becoming the Phoenix rising from a thick pile of ashes. America was still reeling from Richard Nixon's resignation on August 8th, and the Vietnam War was finally winding down.

The year 1974 had not been a good time for the arts. Beverly Hills clothing designer Jim Reva told me, "We're in a depression. At least in my business." And, it wasn't just his business. Money was very tight, which made financing films far more difficult than in the past.

On the upside, love was still free, hopes were still high, and I, like the youth of America, still believed that the future held infinite possibilities.

CHAPTER 26

OPEN AND CLOSE, OPEN AND CLOSE

Wednesday, December 4, 1974

Unbelievable day. Started early to make my appointment on time (in town). Pouring rain when steering went out of control and I spun across P.C.H. several times—landing squarely against a concrete fence that saved me from going down a sheer embankment. It was a true miracle that I didn't get killed, hit another car, or go flying over the embankment and break my neck.

It all happened within the blink of an eye. A hard rain was coming down, and water had begun pooling, loosening layers of oil that had been accumulating for weeks, maybe, even months. In places it was as slippery as black ice.

I knew it would be slow going, so I left early, heading into town. My agent had set up a meeting with a filmmaker who was looking for a young screenwriter to rewrite a script, and I was excited to be back in the game. Traffic was light since most of Malibu was still in bed or drinking coffee and reading the *L.A. Times*.

I vividly remember stepping on the brake and losing control—sliding across all four lanes of Pacific Coast Highway before spinning around and crashing, rear first, into a concrete wall.

Stunned, I got out to assess the damage, and was horrified to realize that there was a sheer twenty-foot drop to the beach below. If the barrier hadn't been there, I would've flown over the edge and, possibly, broken my neck. Walking away from this mishap was a miracle, an affirmation, and as shaken as I was, I was grateful to be alive and in one piece.

I was also relieved to see the homeowner rush outside to investigate the loud noise. He calmed me down, and let me use his phone to call Jose, at my father's office. Jose was my father's trusty mechanic, and over the years he'd rescued me countless times from my father's hand-me-down vehicles. I had to force myself to keep going. I had to. I had to keep moving forward. There was no other choice.

> Wednesday, December 4, 1974
>
> Jose came out and put gear in park. The car started and I made my appointment. It went very well—nice young filmmaker. He thought my experiencing "business" was an asset.
>
> Went by SOH [School of Holography] and learned that they're about to go under. I suggested a benefit which, amazingly enough, brought Cecile and I together. [capable of working together]
>
> [That night] In the kitchen playing the Dictionary Game with Jay, Deuce, Mary, Alfred, Cassie and Gino—past 7:30 and I'm giving up [waiting for Paul] when—wap! In walks Paul [Whitehead]. He brings his spellbinding paintings into my bathroom and we have a look at the Zen Collection—"Bull" Series of 10.

Even though the rest of the day had gone well, I was still shaken from the accident. The station wagon had survived with a few bumps and dents. I, on the other hand, was sore and depressed. I needed something to go right.

I was about to call it a night when Paul Whitehead arrived, umbrella in hand, and finished artwork under protective cover. He was hyper-focused, anticipating his first Los Angeles art exhibition. What did I think? What did Cassie think? Would we invite collectors to his opening?

As we admired his "Bull" series, raindrops could be seen dripping, first in the living room, then the kitchen. It seemed that every room, with one exception, leaked.

"Save my paintings!" Cassie screamed, and off we went, rushing from one room to the next, pulling her paintings off the walls. A flurry of teamwork found us branching out to our own leaky rooms to save our own treasures. On Cliffside Drive there was always an element of drama mixed with exhilaration.

A few nights later, Paul Whitehead wined and dined me at the Sand Castle, our local waterfront eatery. Sitting in a banquette and staring out to sea, Paul made it clear that he didn't want to have a serious relationship. He was ambitious, and I was a good door opener. Disappointing as this might seem, at least he was honest. Without my even realizing it, he'd pinched my Achilles heel, encouraging me to mistake his attention for an emotional connection.

Ever the optimist, I thought I could win him over by introducing him to people who could propel his career. I took him to meet Paula and Malcolm Cecil. Malcolm showed him Tonto, his room-size synthesizer, and the two Englishmen hit it off. Paul was well known in England for his surrealistic album covers, and with gold records lining Malcolm's walls, he couldn't help but value this introduction as a stepping stone to L.A.-based record companies. Paul was single-minded. He had a show opening in three weeks, and he needed it to be a success.

I took him to meet my friend Lynn Weston in Beverly Hills, and to dinner with my attorney at the Luau in Beverly Hills. Paul asked me if I'd like to co-author his novel. "I have nine chapters finished," he prompted.

Wednesday, December 11, 1974

[Paul and I] Discussed [his] book as we drove into town so Paul could be interviewed by Barbara Birdfeather at the Hollywood Daily News. I'm giving Paul the most coveted kimono for Christmas. I know he'll love it. I think he's beginning to see me as I am, an old fashioned type instead of a "Hollywood" type. The girl he lived with for three years made him feel guilty for painting

> at night. Like Bill the first man I lived with) making me feel guilty for writing at night.

Paul thought it would be fantastic if we could get a "movie bull"—a bull that wouldn't skewer anyone—for his art opening. I thought of Jenny Arness, the lovely, complicated twenty- four-year-old daughter of James Arness, the star of TV's *Gunsmoke*.

Jenny was the mysterious tenant living in the apartment above the garage on Cliffside Drive. She had her own kitchen so there was little reason for her to make the trek to the main house. Oh, she'd pop in when she felt like socializing, but most of the time she kept to herself. She and her ex-boyfriend, Gregg Allman of the Allman Brothers Band, had what could be called a passionate, tumultuous relationship, and even though they'd split up, the rocker continued to visit her when he was in town.

In my diary I describe Jenny, the tall, thin actress with long brown hair, saying: "She's very mature in the 'paid my dues' department." Jenny wanted to act, but even more than that she wanted Gregg Allman back. She suspected that his dalliance with Cher was more than a rock 'n' roll fling.

I introduced her to Paul. Impressed with his "Bull" series, she promised to call her father's business manager to find out if a movie bull could be appropriated for his opening.

On another day, I took Paul to the School of Holography to see how holograms were made. I thought the limitations he'd placed on our relationship had been replaced by a duality of purpose, but he continued to give me mixed signals. Sometimes I thought we were in sync, and sometimes I just felt used.

> Friday, December 13, 1974
>
> He told me he picked up a hitchhiker who said, "You look familiar..." Finally, they realized that they'd met briefly at the School of Holography [in Santa Monica] and Vic was on his way to the School of Holography in San Francisco. "Do you know Vic?" I've met him a couple times." "Well, he knows you." Whatever that

means. It seems that everyone talks about me. I left Jupiter 5 and suddenly I'm a horrible person because I'm living in luxury.

More or less secured an udder-less black longhorn movie cow who plays bulls for Paul's [art] opening.

Shortly after our trip to the School of Holography, Paul left for Los Gatos, just south of San Francisco, where he'd been living before moving to Los Angeles. According to him, he'd picked up a hitchhiker from the San Francisco School of Holography who knew me from the Santa Monica SOH. Thanks to Cecile, Vic—the holographer—had heard only her side of the Jupiter 5 demise.

Jupiter 5 was supposed to be our golden ticket. It was supposed to be the foundation upon which we, the futurists, would continue to change and improve the world using lasers and holograms. Had our original mission been scuttled? What had begun as a glorious dream had turned into a debilitating nightmare.

Cecile and I had become full-on adversaries. She had made the School of Holography in Santa Monica her home base, and not unlike Jupiter 5, it had run out of money. She couldn't cover the overhead, let alone pay instructors.

My new Malibu address convinced her, and in turn, her acolytes, that I had received insurance money from my father, and had moved on. Cecile aired her grievances, and the holographers listened. Yes, I had moved on. Yes, I was living in a fabulous house in Malibu, but no, my father had not given me one cent of insurance money.

It was easy for them to brand me a dilettante and a villain. I've never believed in burning bridges, but sometimes, as in this case, there was no other way to survive.

I went back to MGM, and told the Shellys how much I needed to make each month in order to pay my rent. If they wanted me to write for them, they had to step up. One-third of nothing is still nothing.

Even after I'd leaped to the wrong conclusion about "SOS" in *W.W. and the Dixie Dancekings*, I'd kept the bridge between John Avildsen and I, open. I'd learned my lesson, and I wouldn't make that mistake again.

I had Now Productions send him a copy of *The Laughingly Lurid Life of Daddy and Peaches*. He said he really liked it, and a meeting was set.

> Monday, December 16, 1974, 1 PM Polo Lounge
>
> Incredible day! Had terrible butterflies fearing the John Avildsen lunch. It was all for naught. Lunch went well. John wants to direct Daddy and Peaches. He has a "cold" aura as Shelly (Davis) noted. He's not easy to get close to. I did and it didn't work. The meeting was a success. Shellys observed that they'll all hate each other by the end of the film.
>
> 3 PM back to the studio, and by 4:40 had completed treatment for Off the Wall, and had it delivered to Joe Goodson by 5. His office is right out of Sloanes' window (Beverly Hills department store on Wilshire). Posh, in Thalberg Building. He loves the idea!

The two Shellys and I met John Avildsen at the Polo Lounge at the Beverly Hills Hotel. Seated in Shelly Davis' favorite banquette, I was cautiously optimistic. The director was keen and cool, but by the end of the lunch he'd expressed an interest in directing *Daddy & Peaches*. This was huge! John Avildsen wouldn't say he was interested in a project unless he meant it.

Fortunately for John, and unfortunately for us, another project called *Rocky*, a little movie starring a virtual unknown called Sylvester Stallone, swooped in and captured the director's attention. Our dark comedy disappeared into his rear-view mirror.

> Tuesday, December 24, 1974
>
> 10 AM—Phone rings and it's Milton Moore. "I have $500 in my hand and I want $500 worth of Lumilar."
>
> I have to call Cecile. Yes, I can have it at Jupiter 5 cost and make a commission. She won't give too much. "I may be going into town later, so four wouldn't be good. How's one?!?"

"All right, make it one," I said.

I knew it would be difficult, but I had no concept of the vitriol awaiting me. Cecile was sitting at the bar counter separating the kitchen from the living room. She didn't budge. Her eyes were cool. She'd thought everything out and discussed it with "her boys." She didn't want to sell any of her supply, but I insisted that you can't ask someone to come up from Huntington Beach expecting to buy $500 worth of material and get nothing. She finally conceded, bringing out four rolls of silver with adhesive. I could take 21' of each—first roll she stopped at 18'). I was furious, but my hands were tied. Rod lounged on the other side of the counter, drinking out of my coffee cups, not lifting a finger to help me measure, then making me pay $5 to get my own portfolio back, broken and with none of the news clippings I'd so carefully saved. "You can make photocopies," she said as she snatched them out. I wanted to punch her, but that would only make things worse... The worst: I'm measuring 3/4"M—18' "That's it!" Barney Kaelin walks in—

"Oh, can I buy some of that?" Cecile says, "Sure, $5.40 per yard." "All I have is $5." "Oh, that's cool—it's Christmas."

First National Bank still wanted the money they'd released when I cashed Richard Aaron's bad check. At first, the banker suggested I take out a loan, and then, when he realized I had no collateral, he insisted I pay the debt back $50 a month.

At this point I felt like I'd repaid that money a thousand times over.

Wednesday, December 25, 1974, Christmas

Sat in the sun, got it together, and went to parents for, hopefully, the last round of bad, 30 years of Christmas. I walked in and heard, "Change that top, it's terrible!" I went up and changed clothes.

> "Let's open presents." "No, not now, the people will be here any minute." And, so it went, the tug of emotional wars. Mother screaming at me and vice versa. Herb and Joanne arrived with one dozen frozen string cheeses, then the Sinnotts. Next, we're opening gifts and I get a top, a small bottle of Replique. "Daddy, where's mine from you?" He looks sheepish, says nothing and keeps opening his gifts. Finally, he says, "I thought your mother bought you a lot of presents." I'm overwhelmed, overwrought. I quietly go into the bar and cry. Daddy comes in with the L.A. Times Bullock's Wilshire half-price sale page. I continued to cry. Later I went to my room. Mother brought her Xmas present to herself, a solid gold thimble with sapphires. "Someday this will be yours." I cried harder. Don and Rosy Rogers think I'm an ungrateful brat because I kept making nasty cracks. It's an irreparable situation.

I always dreaded spending Christmas at my parents' house. 555 North Bristol Avenue, as I've said before, had a strong stamp of Los Angeles approval. The house owned us, not the other way around. And, my parents were usually great hosts—in fact, they were always better with guests than family.

Every morning began with bickering and sometimes a full-on argument. Christmas heightened the stakes. My mother was never happy with anything my father or I gave her, and my father always received so many nice gifts from business associates, he didn't really mind what we gave him.

Why, you might ask, did I keep going back? I've given it a lot of thought, and realized that, like someone with Stockholm Syndrome, I was a willing moth to a bonfire.

As an only child with a lifetime of psychological, and occasional physical abuse, I was as well-trained as the Springer Spaniels. And, like the dogs, sometimes I acted up and had to accept the punishment. Ironically, my parents came across as convivial pillars of the community, leading their friends to believe that all my problems were of my own making. Outsiders had no idea what went on behind closed doors.

The residual effect of so much toxicity was a lack of self-esteem. In order to deal with day-to-day life, I would flip a mental switch that would allow me to play a scene in my own movie. I appeared to be living in the moment, but I was really shielding myself from that moment, using a coping mechanism. It wasn't until the end of the 1970s that I realized that this feeling, or the ability to step outside of oneself, was not a universal experience.

One night I asked screenwriter Ivan Moffat, "Don't you feel like you're playing a scene in a movie?" He scowled and shook his head. "No." His eyes narrowed and he looked at me with concern. "No, Linda, no one feels that way. I can assure you."

Ivan was right. I needed to find my center, so to speak. I needed to live in the moment, not in the past or the future, and to do that I needed professional help in the form of therapy, yoga, meditation, and truth-saying, supportive friends.

I learned that being an optimist was important, but it wasn't enough. I still had to rely on my coping mechanisms to help me navigate the Hollywood landscape. And, it wasn't easy since the friendship, for me, always seemed to be conditional. One day you were best friends, sharing intimate secrets, and the next, you barely acknowledged one another. I found it to be a culture that rewarded beautiful, thick-skinned individuals.

If I had to give my younger self one piece of advice it would be this: Pick one thing and stick with it. I had too many possibilities and all of them were contingent upon other people.

Slogging through one's past can be tedious, however. Having daily diaries from the early 1970's, with detailed descriptions was enough to send memories flooding through me. I was forced to relive the good, the bad, and the ugly.

Tuesday, December 31, 1974

As the tide rolls back and forth beneath our castle by the sea, the shape of '74 is laid out in brilliant sequence. Beginning slowly,

> being told that if you haven't made it by thirty, you're washed up. Making it in the futuristic marketplace, Star Trekking our way to a local position of prominence, wondering at the crushing pace, yet so caught by the power of the wave that there was no question as to riding it through. Reaching peaks or climbs that sparked radiant highs.
>
> All the energy expended. So much it makes you sigh. Then meeting all the con artists—each one smelling success, each one biting and sucking until the well was dry. Bobbing for air when none seemed available, racing all over town trying to go to Hawaii, then making it First Class. Compression. The accident, the party, Julia, Malibu, gypsying it for eight months. Searching for the "right" man, rejecting all others no matter the circumstance. Falling into my dream house at Point Dume featuring the loonies, the wonderful cast of drama kings and queens. Then Paul and all the companionship of a perfect bond, yet one dynamic is held in abeyance. In matters of the heart, reason goes out to sea. The foundation is laid, now the walls must be finished. This is a time of solitude, rest after battle. My heart is light as I hope for the best.

Here's my takeaway: In spite of so many challenges, I survived. I went on to publish books, write for film, and produce fine art holography exhibitions. My vision of the future back in those heady, chaotic, 1970s Hollywood times has proven itself to be prescient. Holograms are now mainstream. They appear on our credit cards and are woven into new hundred dollar bills. Larger-than-life images of dead superstars—Whitney Houston, Michael Jackson, and Elvis Presley—appear on stage to give live performances. I still love holograms, and I still believe that they will continue to change and improve our lives.

EPILOGUE

"Hey, Linda, did you hear? Your roommate killed herself."

"What?"

On May 13, 1975, I walked into Now Productions at MGM, and these were the first words out of Shelly Davis' mouth. He carried on, saying, "I heard it on the radio."

My roommate, Jenny Arness, had been living in the guest apartment over the garage on Cliffside Drive. Her father was James Arness, the star of *Gunsmoke*, one of America's most popular television programs, hence the immediate media coverage.

What followed was a nightmare to punctuate our grief. Paparazzi staked out the street and brazenly trespassed to take pictures of the remaining residents sitting on the cliff eating breakfast.

My dear friend Peter Cookson had flown out from New York and happened to be visiting when the Indian tribe Cassie hired to exorcize the property arrived. Peter viewed my situation going from very bad to much worse. "You've got to get out of here!" he insisted. "You need to be in L.A." And with that, he moved me into the Chateau Marmont, which proved to be the beginning of a new life.

I was hired to rewrite Harlan Ellison's screenplay for *I Robot* for Warner Brothers, and to join the Advisory Board of the Museum of Holography in New York City. In 1976, with Candy Clark's help, I talked producer Si Litvinoff and director Nicolas Roeg into using a multiplex hologram of David Bowie in The Man Who Fell to Earth.

Many holography-related accomplishments followed. I consider my most significant contribution as forming Holographic Management Associates with Danna Ruscha and Randi Foster, to produce the first fine art holography exhibition in Los Angeles.

I had pounded the pavement along La Cienega Boulevard, the gallery row of its day, asking every gallerist if they would sponsor an exhibition of holograms. They all said no. Finally Molly Barnes and Jackie Anhalt at the Anhalt-Barnes Gallery agreed to take a chance.

Molly Barnes was especially passionate about emerging artists and innovators like Bruce Nauman, who were making holograms that had started to attract collectors. Molly and her partner, Jackie, were skeptical. They didn't know if there was a market for three-dimensional, hard-to-light and hard-to-hang art. They offered us a deal: five-hundred dollars to rent the gallery for one month, plus a small commission.

Being in the heart of Los Angeles' highly respected art scene gave Holographic Management Associates credibility. I had recently been paid for my work on *I Robot*, and knowing that artist-holographer Rudie Berkhout would only let you show his work if you purchased a piece, I bought two limited edition white-light transmission holograms. For every hologram you purchased, he would loan you one. We were able to show four pieces of his masterwork.

"Holography '79, Series I" sold out. The show was a rousing success. The question on collectors' minds was: Will these images hold up over time? The answer to that question continues to unfold forty years later.

The following year, HMA produced "Holography '79 Series II" at the Anhalt-Barnes Gallery. I can remember showing author Michael Crichton a beautiful Stephen Benton reflection hologram that was suspended from the ceiling. Crichton was very tall and he couldn't figure out how he could hang the artwork in his house so that people of varying heights could see it. At that point in time, I didn't have a solution.

It was during this period that I made a deal with producers Si Litvinoff and Harry Blum for *Laser Lady*, my original screenplay. Si had been the executive producer of *A Clockwork Orange* and Harry Blum had produced a number of feature films that starred notables such as Elizabeth Taylor and Jane Fonda. Si and Harry optioned the rights to *Laser Lady* and made presentations at the Cannes Film Festival to raise money. I met with a young actress named Sissy Spacek who had become Hollywood gold after her performance in *Carrie*. We thought she would make a great Laser Lady.

I flew to New York to interview Salvador Dalí in the King Cole Bar at the St. Regis Hotel. Eccentric as Dalí was, he loved visual magic. I gave him a dichromate hologram and he refused. "I don't take gifts!" he said emphatically. Then ten minutes later he snatched it away from me.

It was the end of the 1970s, and I was ready to make my very own blockbuster. That is until early one Sunday morning when I got an alarming phone call from my father. As a small child, he had taken great joy in sharing the Sunday comics with me. Apparently, he still read them. It seemed that Disney had a comic book promo for a new film called *Condorman* in the Sunday *Times*, featuring a character called "Laser Lady." She even looked like the artwork that had been circulating for my upcoming film.

One might wonder why my father would call me about this when we had such a contentious relationship. Whether he approved or not, he knew that I'd spent years creating this character and writing an original screenplay, and he respected those efforts. Back in the day, as a contractor, he'd bid on a unique movie theater in Hollywood. He'd been the only contractor who'd been able to tell the developers how they could actualize the architect's rendering. In the end, they gave the job to someone else. My father never got over the injustice. On principle, he wasn't going to allow Disney to steal his daughter's creation.

Condorman, the film (1981), was based on a book, and "Laser Lady" was not in that book. My producers' option was about to expire, so Litvinoff and Blum told me I needed to sue Disney.

I don't recommend litigation to anyone. It's expensive, and often, corporate deep pockets dictate how long the legal proceedings will continue. My battle with Disney forced me to address the David and Goliath nature of our justice system.

To date, I've produced two issues of *Laser Lady* comic books, based on my original screenplay. Singer/songwriter Clive Kennedy and I used my screenplay to map out a musical, while producer Willette Klausner has spent decades pitching *Laser Lady* for film and television. As always, for me hope springs eternal.

NOTABLE PEOPLE IN LASER LADY MEETS THE LIGHT JUNKIES

Lou **Adler,** record producer

Steven **Arnold,** avant-garde photographer and artist

Hal **Ashby,** film director

John **Avildsen,** film director

Tom **Baker,** actor in Andy Warhol films

Rona **Barrett,** Hollywood gossip columnist

Jerry **Brandt,** agent and producer

Sheldon **Brodsky,** agent and producer

Jody **Burns,** original member of Holographic Communications Corporation of America

Jack **Calmes,** inventor, musician, founder of Showco

Dyan **Cannon,** actor and director

Malcolm **Cecil,** English musician, record producer, recording engineer, electronic music pioneer

Candy **Clark,** actress and model *(American Graffiti)*

Michael **Clark,** Byrds drummer

Peter **Cookson,** producer and entrepreneur

Alice **Cooper,** musician

Francis Ford **Coppola,** filmmaker

Michael **Crichton,** author

Lloyd **Cross,** physicist, holography pioneer and founder of San Francisco's School of Holography

Salvadore **Dalí,** surrealist artist

Sheldon **Davis,** co-founder of The Whiskey a Go Go, producer

Ivan **Dryer,** father of the laser light industry, co-creator of Laserium

Doris **Duke,** tobacco heiress and art collector

Rocky **Dzidzornu,** Ghanaian-born English percussionist.

Robert **Easton,** actor and recording artist

Freddie **Fields,** agent, producer, and studio executive

Michael **Foster,** holography pioneer

Elsa **Garmire,** physist, laser artist, co-founder of Laserium

Rudi **Gernreich,** fashion designer

Bob **Gilbert,** artist, filmmaker, and builder

Barry **Gott,** artist who specialized in light boxes

Richard **Harris,** actor, singer, and writer

Don **Ho,** popular Hawaiian entertainer

Margot **Kidder,** actor

Lynn **Lenau,** filmmaker

Selwyn **Lissack,** artist, musician; created holograms for Salvadore Dalí

Donyale **Luna,** supermodel

Si **Litvinoff,** film producer

Peter **MacGregor-Scott,** film producer

Bob **Margouleff,** record producer, recording engineer, electronic music pioneer

Gertrude Ross **Marks,** Golden Globe-winning writer and documentary film producer

Walter **Matthau,** actor

Kim **Milford,** actor, singer-songwriter, and composer *(Jesus Christ Superstar)*

Ivan **Moffat,** screenwriter

Chip **Monck,** Tony Award-nominated lighting designer; master of ceremonies at the 1969 Woodstock Festival

Lon **Moore,** holographic artist, first student at the School of Holography

Ana Maria **Nicholson,** early influential holographic artist

Joan **Nielsen,** artist and writer

Anaïs **Nin,** writer

Charlie **Patton,** A member of Elvis Presley's "Memphis Mafia"

Dennis **Pelletier,** aka Amir Façade, diffraction grating artist for Jupiter 5

Jerry **Pethick,** Canadian contemporary artist and sculptor

Ardison **Phillips,** artist, restaurenteur, winemaker

Roman **Polanski,** film director

Carlo **Ponti,** Italian film producer and husband of Sofia Loren

Nicolas **Roeg,** film director

Cecile **Ruchin,** original member of Holographic Communications Corporation of America

Ed **Ruscha,** artist

Leon **Russell,** musician

Red **Shepard,** actor, singer, and producer

Steven **Solberg,** artist

Ringo **Starr,** Beatles drummer

Dean **Stockwell,** veteran film actor

Ann **Turkel,** actor

Lew **Wasserman,** head of MCA, a media conglomerate that merged with Universal Studios to become MCA Universal

Kaisik **Wong,** clothing designer favored by Salvador Dalí

Dr. Ralph **Wueker,** pioneer in laser physics and holography

ACKNOWLEDGMENTS

Forty-eight-years ago I stepped into the world of lasers and holograms. It was an extraordinary time with awe-inspiring opportunities. For me, the early 1970's marked a coming of age, a transition, and a forced maturation.

Being so personal, my diary recounted so many escapades, embarrassing moments, and liaisons that I was forced to ask – what will people think? "Who cares?" was the consensus. My family and friends kept reminding me that my daily diaries reflected a specific point in time, a time when possibilities were infinite and living each moment to the fullest fueled our actions.

First, I must thank my daughter, Lucy, for finding holograms as magical as I do. As a child she was fortunate to grow up around both Hollywood icons, artists and physicists who were pioneers of holography. The confluence of the two worlds stimulated her imagination, inspiring her to focus on neuroscience. Lucy's husband, Philip, is my favorite sounding board. I value Lucy and Phil's insights, support, and suggestions. They improve life itself.

I began this memoir several times, but it was only when my longtime friend, Stephanie Ricardo began reading and commenting on chapters that I gained the momentum to keep going forward and, ultimately, finish the book.

Delving into one's past, especially an emotionally tattered childhood, made having a sincere support system essential. For that, I must thank Robin and Jim Hollister. Both avid readers, their encouragement motivated me to write about my adventures. Their support helped me to be open and honest.

Candy Clark has been a great and caring friend who will always tell me the truth. I value her opinion above most.

To Clive Kennedy, the voice of reason in an unreasonable world, and to the friends who gave me continuous input and support including John, Charlie, Ellen, Julie, Margaret, and Valerie.

I'd like to thank all of the holographers who knowingly and unknowingly contributed to the early history of holography. Salvador Dalí may have been the first major artist to use the medium, but it is Peter and Ana Maria Nicholson who turned a novelty into fine art. Their holograms opened museum doors and appealed to collectors around the world.

I am indebted to my editor Maya Ziobro, designer Sarah Clarehart, and publisher Michael Roney for graciously helping me to bring this chapter of my life to a close. Working with them has been a wonderful, seamless experience.

INDEX

Note: Page numbers in *italic* indicate figures.

D

I

J

K

L

M

N

O

P

R

S

T

V

W

Y

CPSIA information can be obtained
at www.ICGtesting.com
Printed in the USA
BVHW010919200720
584122BV00008B/49/J

9 780